Robert M. Miller Communcations

Imprinting and Early Learning

for the Newborn Foal

A highly effective method for shaping a horse's lifetime behavior

By Robert M. Miller, D.V.M.

**Imprint and Early Learning
for the newborn foal**

*Published by Robert M. Miller Communications
14415 Donnington Lane
Truckee CA 96161
www.robertmmiller.com*

Cover painting of Robert M. Miller, D.VM.
By Dwayne Brech

Back Cover Photography by
Debby Miller

*©Copyright 2023 by Robert M. Miller Communications
All rights reserved. No part of this book may be reproduced in any form by any means, electronic or mechanical, including photocopying and recording, or any information storage or retrieval system, without permission in writing from Robert M. Miller Communications.*

*Originally published by Western Horseman Magazine
as Imprint Training of the New Born Foal*

ISBN 978-0-9834625-7-6

DEDICATION

My favorite task during 32 years of practice was being called to attend a newborn foal. I never tired of seeing the miracle of birth and the beautiful creatures that birth produced. New foals have given me great pleasure over the years. It is a joy to watch them and I learned to predict, with considerable accuracy, their future potential. With the hope that I can make the lives of foals a little easier and a little more pleasant,
I dedicate this book to them.

ROBERT M. MILLER, DVM.

"Do you not know, then, that the beginning in every task is the chief thing, especially for any creature that is young and tender? For it is then that it is best molded and takes the impression that one wishes to stamp upon it."

Plato
The Third Century, B.C.

ACKNOWLEDGMENT

My interest in equine behavior, and my proficiency in modifying and shaping behavior in horses were enhanced by authors-some living, and some long gone. Among the most influential were D. Magner, Horace Hayes, and Kell B. Jeffery, Tom Roberts, Dr. Jim McCall, all of whom have passed on. Still living, and still influential, are Desmond Morris, Dr. Sharon Cregier, Dr. Katherine Houpt, and many others who write about equine psychology.

I am also indebted to the scores of fine horsemen and horsewomen who have given me additional insight into the behavior of the equidae. These most notably include Pat Parelli, Ray Hunt and Maurice Wright.

Horsewoman and English teacher. Laura Wade was very helpful in preparing this manuscript. Special thanks to Dr. Donald D. Draper for his invaluable technical assistance. I must also express my gratitude to my wife, Deborah, for her keen observation skills and ability to interpret the responses she sees in horses, and for taking many of the photographs for this book.

Additions

Dr. Miller recommends the following during the imprint training procedure: When the mare stands on her own after giving birth, milk 4 to 6 ounces of colostrum from her and give it to the new born foal via a baby bottle. According to recent research, this reduces chances for septic foal syndrome, one of the most common causes of foal death.

Other than receiving this vital dose of colostrum, there's no urgency for the foal to nurse. It's safe to delay further nursing until you've completed the imprint training procedure.

INTRODUCTION

This book was written for the working horseman, rather than for the formally trained behavioral scientist, psychologist, or ethologist. It will be obvious to the professional behaviorists, many of whom have published works which I have avidly read and who have contributed greatly to my understanding of the subject, that I lack a formal academic background in behavioral science. I have, however, had a uniquely rich experience in the animal world.

Before graduating from veterinary school, I worked as an employee in a pet shop and a kennel, a milker on a dairy farm, a teamster driving draft horses, a veterinary assistant, a horse wrangler, a packer in the Colorado Rockies, on the racetrack, and in the rodeo arena. At various times I was a rodeo contestant, a horse breaker, a cowhand, and an employee at a dog pound.

I went into the United States Army during World War II, and those 2 years were the only time in my youth that I did not work with animals. In 1951 I received a bachelor's degree in animal husbandry from the University of Arizona College of Agriculture, and in 1956 I reached my career goal when I received a doctor of veterinary medicine degree from Colorado State University.

I have always enjoyed working with foals

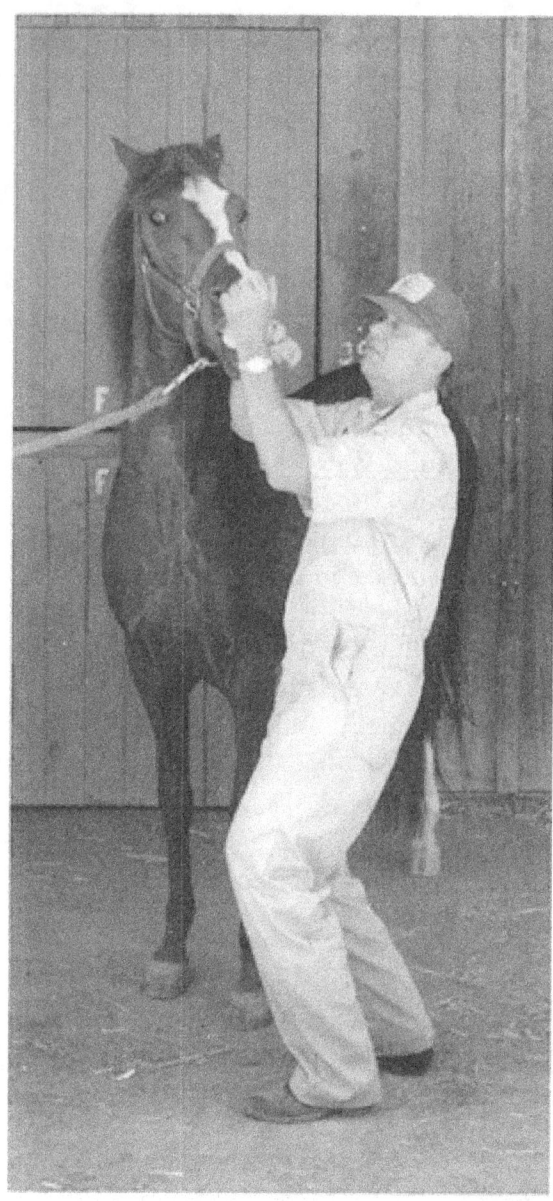

During my veterinary practice, I worked primarily with horses.

During 32 years of veterinary practice, I worked with every conceivable species of domestic animal, and with many zoo species as well, attending to many of the wild animals used in the entertainment industry. I especially worked with horses, and as the years progressed, I spent the largest part of my time with that species. That I enjoy my work is evident in that, together with my family, I have, as a hobby, raised and trained dogs, goats, horses, and mules.

This book could have been more accurately entitled *The Application of Behavioral Shaping Techniques to Neonatal Foals.* Sorry! I have not attempted to write an academic treatise. This is a how-to book for hands-on horse people, especially for those who must work with foals. The dramatic results that can be obtained by the early training of foals should be of vital interest to all horse persons, be they trainers, farriers, veterinarians, riders, breeders, grooms, or scientists. The techniques described are consistently effective if done properly.

After many years of being published by *Western Horseman*, we now go to print on this new edition by Robert M Miller Communications. This book was originally titled *Imprint Training of the Newborn Foal.* This was somewhat confusing because "imprinting" and "training" are different things. "Imprinting" as will be explained later, is a predetermined response, whereas "training" implies behavior shaping and behavior modification methods involving conventional learning responses. We decided to rework the title to address this contradiction, while keeping the term "Imprinting" to aid readers when searching for the title.

Most of the techniques described herein involve learning processes other than imprinting. I am certain, however, that the imprinting phenomenon is involved in the immediate postpartum procedures I describe, and I believe that this imprinting process greatly facilitates the subsequent procedures employed.

I therefore choose to call the entire method imprint training, even though much of it is probably performed after the actual imprinting period. Calling it imprint training furthermore emphasizes the importance of starting training early in the foal's life ... the earlier the better.

Whether the techniques employed after the immediate postpartum period involve classical critical and sensitive learning periods, or simply represent learning by reinforcement in a young and previously unconditioned subject, is an interesting question. I hope that my effort will stimulate academic research in that regard. For the working horseman, however, what is important is that these methods work.

Laying the foundation for teaching a horse most of what he will need to know to serve as a useful animal for the rest of his life can be done before he is 4 days of age. Early training can, in an amazingly brief period of time, assure an ideal relationship between horse and human, with the horse bonded and submissive to the human.

In addition, the horse will be desensitized to the ordinary frightening stimuli which usually elicit a flight reaction in the young horse, and which account for the frequent injuries that afflict horses and the people who work with them.

Examining a friendly lion.

tists to explain why it is so effective.

I hope that my efforts will help to popularize the training of very young foals, because presently, a widespread dogma exists that discourages human contact with the newborn. If I am successful, I will be gratified by the lessening of trauma and an adversarial relationship between horses and humans.

—*Robert M. Miller, D.VM.*

My practice also included zoo animals, such as this chimp who's trying to be helpful.

Moreover, the baby foal can quickly be conditioned to respond to certain physical stimuli, making him easy to lead, to tie, and to move in various directions.

Let me assure the reader that I have no illusion that I have discovered something new. Humans have worked with newborn foals and shaped their behavior (beneficially or adversely) for thousands of years. There is historical evidence to that effect. All I have done is given the procedure a name, ritualized it, and used the growing knowledge provided by behavioral scien-

Meet Dr. Miller

Dr. Robert M. Miller has dedicated his entire professional career to the care of horses—first as a doctor of veterinary medicine and today as an internationally acclaimed author, lecturer and clinician developing and promoting safe, gentle horse-handling techniques. He's at the forefront of the horsemanship revolution that embraces better horse-human relationships through understanding, communication and a natural approach to living with, riding and caring for horses.

After graduating from the University of Arizona with a degree in animal husbandry in 1951 and earning a doctorate in veterinary medicine from Colorado State University in 1956, Dr. Miller founded the Conejo Valley Veterinary Clinic in Thousand Oaks, Calif., and was its chief-of-staff when he retired in 1987.

During that time, he had ample opportunities to observe the nature of horses and how the equine mind works. Through study and experimentation he conceived his theories on equine learning and how early experience shapes and influences the rest of the animal's life. He calls this process "imprint training" because it's the training during the imprinting period, which is right after birth.

Dr. Miller now devotes himself full time to helping others understand the psychology and physiology of horses and to advancing the science of horsemanship. As a sought-after speaker, Dr. Miller has given over 400 lectures or seminars to universities, veterinary associations and animal organizations in North and South America, Europe, Africa, Asia, Australia, New Zealand and the Middle East. His topics cover various aspects of veterinary science, horsemanship, ethology, animal behavior-shaping and veterinary man agement and philosophy. He's acted as a consultant for the pharmaceutical and veterinary supply industries, as well as legal consultant and expert witness.

Just some of Dr. Miller's professional awards include the California Veterinary Medical Association Award of Merit (1973), American Animal Hospital Association Award of Merit (1978) and Award for Outstanding Service (1989), Bustad Companion Animal Veterinarian of the Year (1995) and American Veterinary Medical Association President's Award (1996). Western States Expo Hall of Fame (2004), Western Horseman Award (2011), Western Veterinary Conference Equine Educator of the Year (2012), SCVMA Don Mahan Memorial Award (2013), Alumni.

He lives on a ranch near Thousand Oaks, Calif., with his wife Deborah. His other interests include raising and training horses and mules, skiing, travel, anthropology, writing and cartooning.

Please see the last page of this book for a complete list of Dr. Miller's books and videos

CONTENTS

Pages

6	1/ What Is Imprinting?
13	2/ Bonding
17	3/ Habituation
23	4/ Sensitization
30	5/ Dominance
38	6/ The Mare
42	7/ Immediate Postpartum Procedure
62	8/ The Second Session
72	9/ The Third Session
87	10/ Teaching The Foal To Tie
93	11/ More Halter Training
100	12/ Advanced Halter Training
110	13/ Teaching Perfomance Basics to the Young Foal
114	14/ Reinforcing Responses
120	15/ Preventing Problems
128	16/ The Rice Horse
132	17/ Mules
138	18/ Effects of Imprint Traning on Mares
142	19/ Update Information

1 What Is Imprinting?

IMPRINTING CAN be defined as a learning process occurring soon after birth in which a behavior pattern is established.

The newborn foal, I am convinced, is imprinted to follow and to bond with whatever large object that looms above it at the time of birth. Therein lies the foundation of what I call "imprint training" It is early training, during specific critical learning times, given as soon as possible after the foal is born.

During my youth, while working on western ranches, I learned a lot of horse lore from old-timers. These ideas were

Much horse lore from old-timers was factual, but a lot of it was also completely inaccurate.

In the old days, a colt was usually forcibly restrained and saddled.

Painting by J.N. Swanson

Even though many horsemen believe otherwise, a new born foal can be easily trained.

Vicious stallions, once common, have become a rarity.

passed by word of mouth from generation to generation by men who had spent much of their lives working with horses. Much of this information, of course, was factual, but a lot of it was old wives' tales and completely inaccurate.

For example, I was told that the proper way to start a colt was to "top him out." That is, the colt was forcibly restrained and saddled. A rider got on him and hopefully stayed aboard until the horse quit bucking, after which his "spirit was broken" and he could then be ridden.

That so many horses started in this manner went on to become good, reliable riding horses is more a tribute to the adapt ability of the horse than it is to the method used. No wonder that it was said of so many ranch horses, "It takes a cowboy to ride him". This method is still in use today in some parts of North America and in other areas in the world where horses are raised in large pastures, such as Argentina, Canada, and Australia.

That such methods are unnecessary, crude, and less than ideal has been conclusively demonstrated by horsemen who use gentle, swift, and sophisticated techniques. Some of these horsemen—such as Ray Hunt of Mountain Home, Idaho, Pat Parelli of Clements, Calif., and John Lyons of Parachute, California, Colorado, have made a career of conducting educational clinics on better methods of horsemanship and starting green horses.

Another popular North American myth heard often in my youth was that women should never go near a stallion. It was widely held that the "smell of a woman" could enrage a stallion, and women risked death to be near one. In the latter half of the 20th century, particularly in the United States, women have risen to prominence in the equestrian realm, excelling as riders,

Pat Parelli has become highly successful in teaching people better methods of handling horses. Here, he is demonstrating how he begins to hobble-break a horse.

competitors, breeders, and trainers. In my veterinary practice, by 1980, nearly all stallions of all breeds were being managed by women, thus disproving that myth.

In fact, vicious stallions, once common, have become a rarity. There are several reasons why. One: Probably because stallions, like all horses, are made defensive by fear, but tend to become tractable when subjected to a competent, solicitous attitude such as that shown by most women.

Another myth was the practice of throwing cold water on the rear end of a mare after she was bred; it was believed that this would increase her chances of conception. This silly practice was very common, especially in the western states.

A fourth prevalent myth, at least in North America, was that newborn foals should not be handled. It was said that to do so would interfere with the bonding between mare and foal. I heard that to gentle baby foals was to spoil them and make them obnoxious pets. Besides, most people assumed that since a human baby and a puppy are neurologically immature at birth and cannot be taught much, that the same would hold true of a foal.

Therefore, it was and is still widely believed that newborn foals cannot be trained, and that an attempt to do so would in some way harm the foal. This misconception persists despite documentation, historically, that some horse-oriented cultures did handle and bond with newborn foals.

For example, some Native American tribes were known to do this, as did some Bedouin tribes. Additionally, I have met horse trainers from such geographically diverse nations as Colombia, Germany, and the United States who have told me that it

Imprint training does not destroy the will to run in race horses, even though some horsemen believe that it does.

was a tradition in their families, passed on from one generation to the next, to handle and manipulate foals as soon as they were born.

In the horse racing industry, it is widely believed that to teach young foals good manners is to destroy their flightiness and to therefore destroy their will to run. This fallacy is not held worldwide. An American colleague, a Kentucky veterinarian who spent his first three practice years in Ireland, told me of his surprise to find Thoroughbred foals in that country which were well-mannered and disciplined. The race horse will be more specifically discussed in a later chapter in this book.

Scientific studies have shown that the imprinting and bonding period occurs right after birth and lasts only for an hour or two. After that, the presence of strangers elicits a fear reaction in the foal. It is obvious that this pattern of responses is useful to help the foal to survive in the wild where predators constitute the greatest danger. During that first hour of life, the foal's vision seems to be the primary sense involved in attaching to and wanting to follow large moving objects.

In nature, these objects would normally be the mare and possibly other members of the band. Taking advantage of this, we can simultaneously allow the foal to bond with the mare and with ourselves

Baby foals are fully capable of fleeing danger shortly after birth.

as we go through the procedures described in this book.

I noticed in my early practice years that when I had to assist in the delivery of a foal—manipulating his position in the mare, pulling him out, and then toweling him dry and treating him—that such a foal behaved differently when I saw him again. This was usually at around 3 months of age, when he was vaccinated and given worm medication.

I read every book that I could find on animal behavior, and after becoming familiar with the work of Konrad Lorenz on imprinting in birds, I began to suspect that a similar phenomenon was occurring in these young foals.

My wife and I periodically raise foals, and mules as well. Around 1970, I started to develop a routine which I called "imprint training." Not until I retired from full-time practice did I have the time to fully pursue this interest. I look forward to learning more about the training of newborn foals in the future, because there is much to learn.

My enthusiasm for the method is so great, and I am so eager for horses and people to benefit from it, that I offer this book as a stepping stone. More competent horse trainers than I will adopt this method and use it to achieve things that I am not capable of achieving.

Konrad Lorenz, an Austrian scientist, coined the word "imprinting." Working with newly hatched geese, he observed that the baby goslings were programmed before hatching to attach to and follow the first thing that moved after they emerged from the egg. Usually, of course, this was their mother.

Being thus "imprinted" to bond with and follow the mother and the sibling goslings contributes to the survival of this species.

WHAT IS IMPRINTING?

The little gosling, staying close to his mother and his brothers and sisters, has a better chance of survival in the wild.

Unlike the "ugly duckling" of fiction, the little goose recognized itself as a goose.

Lorenz also learned that if the gosling, when hatched, was exposed not to its mother, but to some other moving object, such as a person's foot or a hand or a dog, that it would attach to and follow that object as it ordinarily would its mother. In other words, the gosling "imprinted" itself on a substitute or surrogate mother.

Lorenz at first believed that the imprinting phenomenon was something that only occurred in birds. Later, it was learned that this process occurs widely in the animal kingdom. It was also learned that there are critical learning periods in the lives of young animals, during which their brains are highly receptive to certain kinds of information.

During these critical learning periods, which are very short in duration, information cannot only be absorbed with great speed, but also with great permanence. Indeed, the information registered during such periods lasts a lifetime and is relatively immune to being changed by subsequent events. In other words, very early in life, permanent attitudes are formed, temperament is affected, and later responses to various stimuli are shaped.

In addition to the critical learning periods, scientists believe that there are also susceptible periods, lasting from many days or weeks (and to months in humans) during which learning ability is greatly enhanced. Obviously, there is much more to be learned, and it is a fertile field for trained researchers.

In many species, the imprinting period is delayed, because at birth the young are neurologically and physically immature. Thus, dogs are born deaf, blind, and helpless, as are the young of cats, bears, birds of prey, and many other animals. In puppies, the socialization period extends from the sixth to the fourteenth week, with the seventh to the twelfth weeks being the most crucial.

Other species, in order to survive in the wild, must be able to detect danger and respond to it immediately after birth.

These are called *precocious* species because, although newborn and small, they are completely developed in many ways. They can see, hear, smell, and run from danger. Baby chicks, ducklings, calves, fawns, and foals are examples of precocious young.

Pertinent to this is an observation I made in Kenya in October of 1989. The topi, one of the many species of antelope common to East Africa, were calving.

They all Topi calve close to the same date, within a period of a few days. The advantage of this is obvious. The predators, presented with such an overwhelming feast, are incapable of killing all the young topi, and enough survive to perpetuate the species. By the second day of life, the calves can run at amazing speed and we saw one outrun a leopard, although we thought that it was injured severely enough in the initial charge that it might have eventually succumbed to its injuries.

We were a veterinary group on a camera safari, and it was interesting to see that the newborn topi were not frightened by our motor vans, although we came extremely close. If, however, we approached a 2- or 3-day-old calf, it would flee in terror. Again, prey species are frequently preprogrammed to attach to and bond with objects seen shortly after birth (the mother, ordinarily, in the wild), and to later flee unfamiliar shapes.

Without such miracles of adaptation, species do not survive in their particular habitat. The environment is a dangerous place for *all* creatures, and it takes complex adaptations of anatomy, physiology, and behavior to offer a chance of survival to reproductive age.

IMPRINT TRAINING

The newborn foal born in the wild must soon get to his feet, recognize and follow his mother, respond to her call, nurse, and run to her when he sees, hears, feels, or smells anything frightening. Above all, he must be able to run along with her at top speed when she flees danger. Flight in the horse is the primary survival behavior, and a few hours after birth, even wobbly foals can run at surprising speed if alarmed.

To summarize: The foal, immediately after being born, can see, hear, feel, and smell almost as well as a mature horse. Infact, I have had to examine and manipulate foals still inside the mare. I noticed that when these foals were born many days later, they did not show the usual fearful response to being touched. I believe that they can become imprinted or, at least, desensitized to the touch of the human hand before birth.

Out of this realization came the concept of imprint training. Using the methods described later in this book can quickly and indelibly teach foals, in the first day or two of their lives, certain vital information. I call this method imprint training, and by means of it, four major goals can be accomplished:

1. Bonding with humans.
2. Desensitization to certain stimuli.
3. Sensitization to other stimuli.
4. Submission to humans.

In the next four chapters, we will consider each of these goals separately and explain how each is achieved.

A newborn foal on the range bonds immediately with his mother.

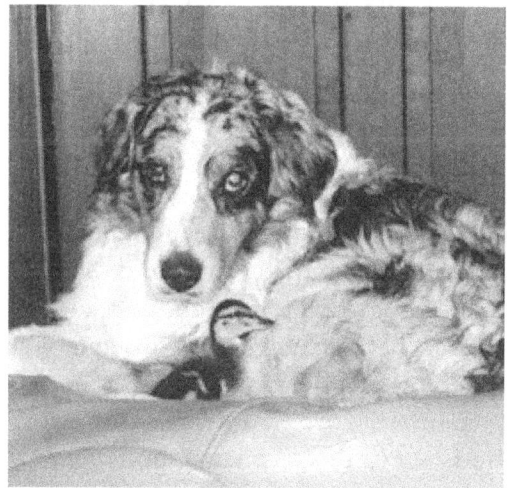

This incubator-hatched mallard duckling first imprinted upon my foot, as that was the first thing he saw when he cracked out of the egg. Later, he imprinted upon my dog, as well.

This is an example of simultaneous imprinting, which is what I believe happens when a foal is imprint-trained at birth; he imprints upon both the mare and handler.

Photo by Debby Miller

2 BONDING

I believe the foal will bond with whatever is moving and above him during the first postpartum hour.

LABOR IS swift and powerful in the equine species, and mares usually lie down to foal. If all goes well, and most of the time it does, the foal is rapidly expelled, but remains attached via its umbilical cord to the placenta, which is still usually within the mare.

Soon, the mare gets up. Additional blood from the placenta then enters the foal's circulatory system through the umbilical cord. When the mare moves around, the cord breaks at a constricted spot close to the foal, and the newborn little horse is now on his own.

Wet and weak, he lies there, busy breathing for the first time. If the mare responds normally, she begins to lick the foal, thus drying, stimulating, and warming him. More important, the taste and smell of the foal arouses the mare's maternal instincts, and she rapidly bonds with him, feeling an overwhelming sense of protectiveness and love.

I believe that the foal is programmed

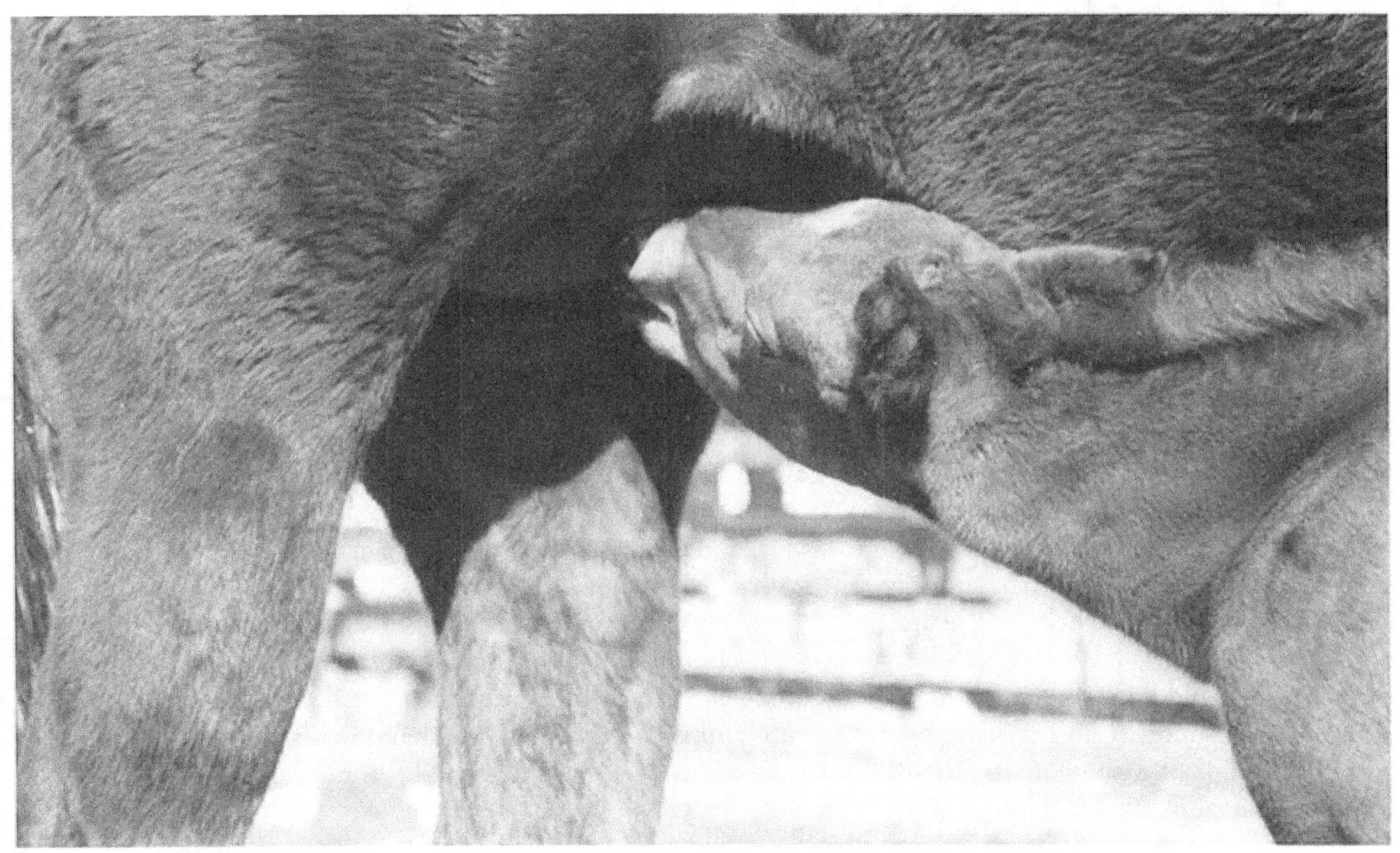

A foal quickly learns where the source of milk is located.

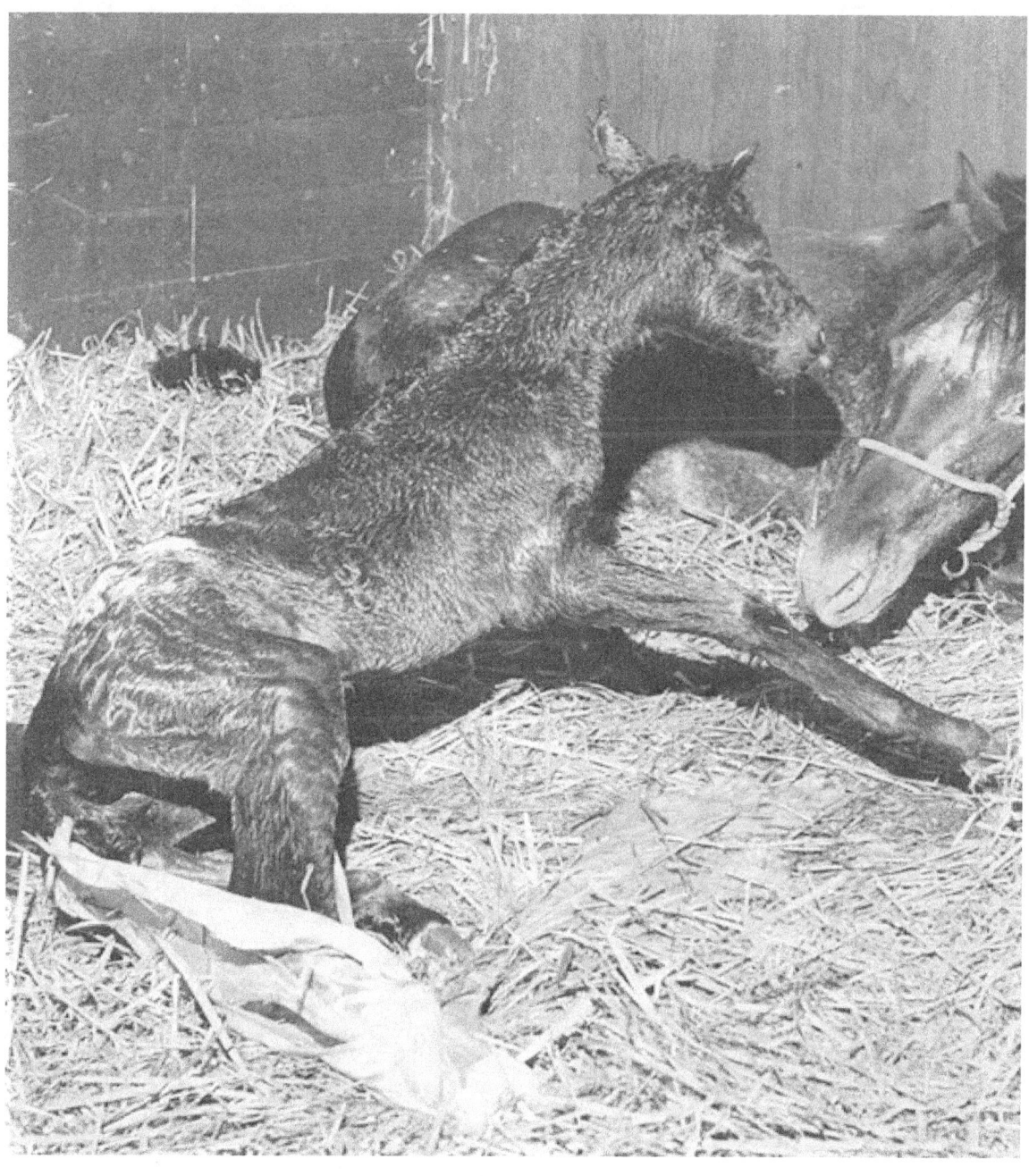

A newborn foal already trying to get up.

Allowing the mare to lick and care for the foal is important.

to attach to and follow with whatever is moving and looms above him during the first postpartum hour. Normally, this is the mare, which is good, because when a foal in the wild is on his feet, he will be imprinted to follow, bond with, and stay close to his mother. This helps to ensure his survival.

In a herd of wild horses, the newborn foal is also exposed to other horses in the herd, and he soon bonds with the herd. Therefore, it is logical to assume that the newborn foal can, as he bonds with his darn, simultaneously bond with other individuals around him, whether they be horses, human beings, dogs, or ducks. This is exactly what can and does occur if a human works with a foal as soon as he is born.

Allowing the mare to lick and care for the foal is important, but if a human, looming above the prone foal just as the mare does, rubs and strokes the foal, and handles his nose and mouth, that foal will be bond-

BONDING

Allowing the mare to lick and care for the foal is important.

Many mares become fiercely protective after foaling. This mare is clearly telling the geldings in the adjacent pen to Back Off! as her brand-new foal (in the doorway behind her) ventures into the corral for the first time.

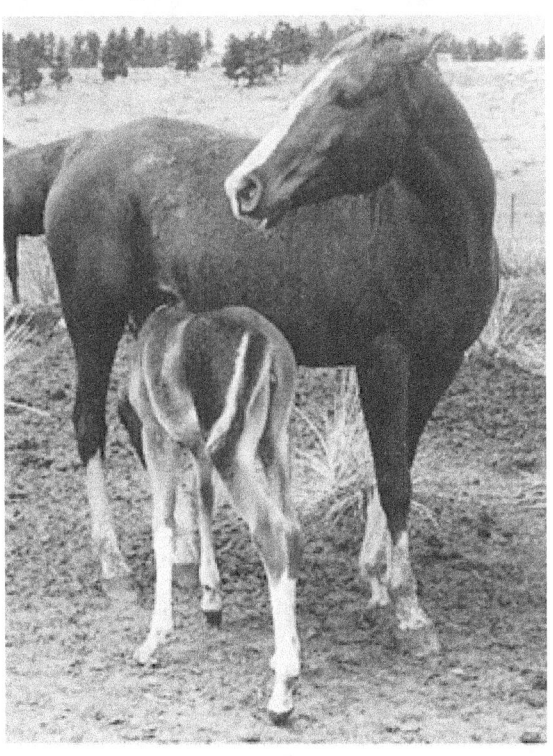
A relatively new mother warning other mares to stay away from her foal.

of milk is, whether it be the mare's udder or, in the case of an agalactic (no milk) mare, from a bottle or bucket. Bonding involves an attachment to the individual, whether it be horse or human, that signifies trust, security, and companionship.

Imprinted foals will often leave their mothers in pasture and come to the person who did the imprinting. This factor alone greatly facilitates subsequent procedures because the *fear* factor is removed at birth, and the foal *wants* to be directed by the trainer.

Normally, under range or wilderness conditions, the foal, a few days after birth, is frightened by any strange movement, unless he identifies the moving creature as one with whom he was bonded right after birth. This fear reaction obviously helps keep foals alive in the wild. If something unfamiliar is seen, he runs from it to the mare to whom he was bonded. Then, if she runs, he keeps up with her wherever she goes.

It should be understood that bonding with a human does not, of itself, remove fear. It simply removes fear of that person, ed with the human just as he is with his mother.

This bonding is independent of feeding. The foal quickly learns where the source just as the foal does not fear his mother.

Otherwise, the foal will respond to unfamiliar sensory stimuli with a fear reaction (flight), whether the stimuli be visual (vision), olfactory (smell), auditory (hearing), or tactile (touch). Suppressing those fears is the subject of the next chapter.

Imprint-trained foals are friendly and readily approach people, even when in pasture.

Horses sometimes develop unlikely bonds-or friendships.
Photo by Karen Tomlin

Even members of species considered natural adversaries can become buddies.

3 HABITUATION

All of us have experienced the phenomenon of habituation.

TECHNICALLY, habituation and *desensitization* are not synonymous terms. Desensitization is a gradual process, wherein habituation refers to the elimination of the response to a stimulus by repeating the same stimulus until there is no longer a response.

In horsemen's language, however, the term desensitization is popularly used to describe both reactions. In either case, the

Show horses must become habituated to crowd noise and other distractions.

The primary instinct of frightened horses is to run.

Although this foal is running because he feels good, he can also run swiftly to flee danger.

natural fearful reaction of the horse to a frightening stimulus is extinguished.

Therefore, in this book, I use the terms interchangeably. If we can eliminate the horse's normal fearful reaction to frightening sounds, sights, and tactile sensations, the end result is the same—and it is a desirable result.

All of us have experienced the phenomenon of habituation. If we sleep in a room with a loudly ticking clock, we ultimately become so used to the sound that we no longer notice it. This is an example of habituation to an auditory stimulus. Similarly, if we take a sea voyage, we eventually get used to the rolling motion of the ship, a tactile stimulus. If we were asleep and the clock should suddenly stop or the ship become motionless, we would probably awaken with a start.

People who work with certain constant odors, as in a gasoline station, stockyards, or factory, become oblivious to those particular odors, because they become habituated to them.

Each species of animal has, as it adapted to its environment, developed emergency survival behavior which helps it to survive. The anatomy of each species is linked closely to this emergency survival behavior. Thus, the armadillo and the tortoise retreat into their armor. The skunk and the porcupine present their weapons in order to survive.

Wolves and other members of the dog family use their ever-ready teeth. The rattlesnake uses its fangs in a different way. The rhinoceros lowers its head, ready to use the formidable weapon nature has equipped it with.

HABITUATION

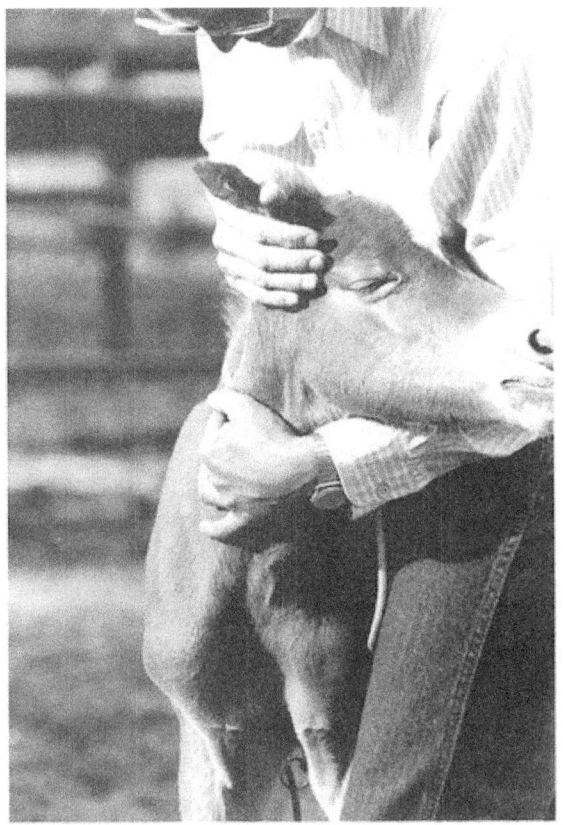

Marty Marten, Lafayette, Colo., is habituating this hours-old foal to having his ears handled.

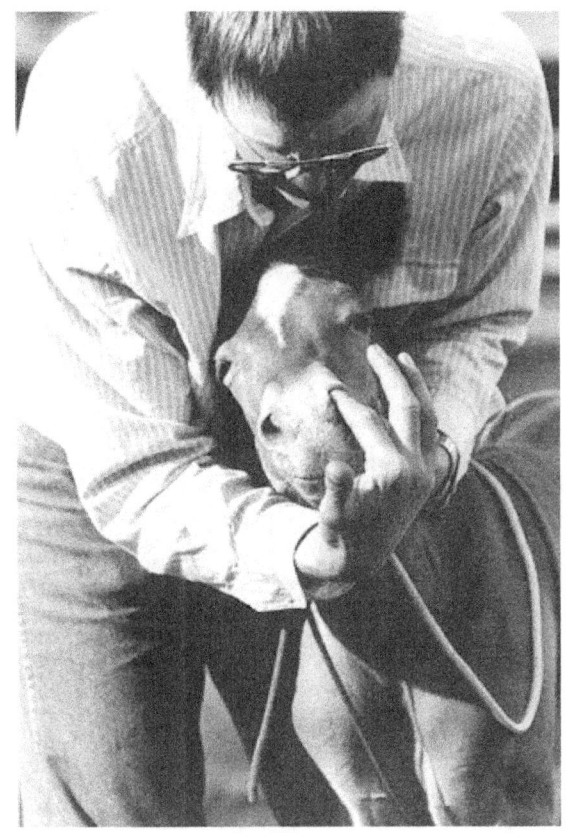

Habituating the foal to the feel of a finger in his nostril. Later, this foal will readily accept passage of a naso-gastric tube for deworming.

Many horses who have never seen cattle are frightened when they are first exposed to them. Putting a horse in a pen next to cattle is a good way of habituating him to them

IMPRINT TRAINING

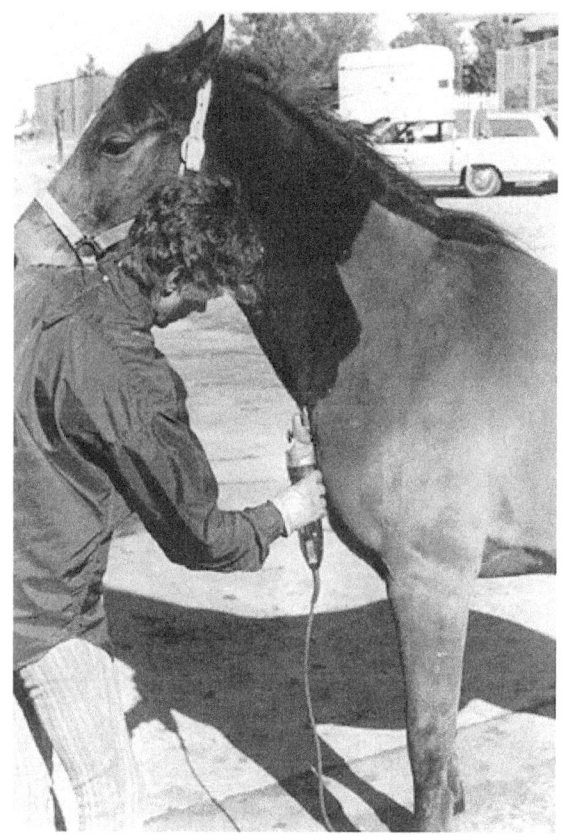

A horse must learn to tolerate the feel of clippers.

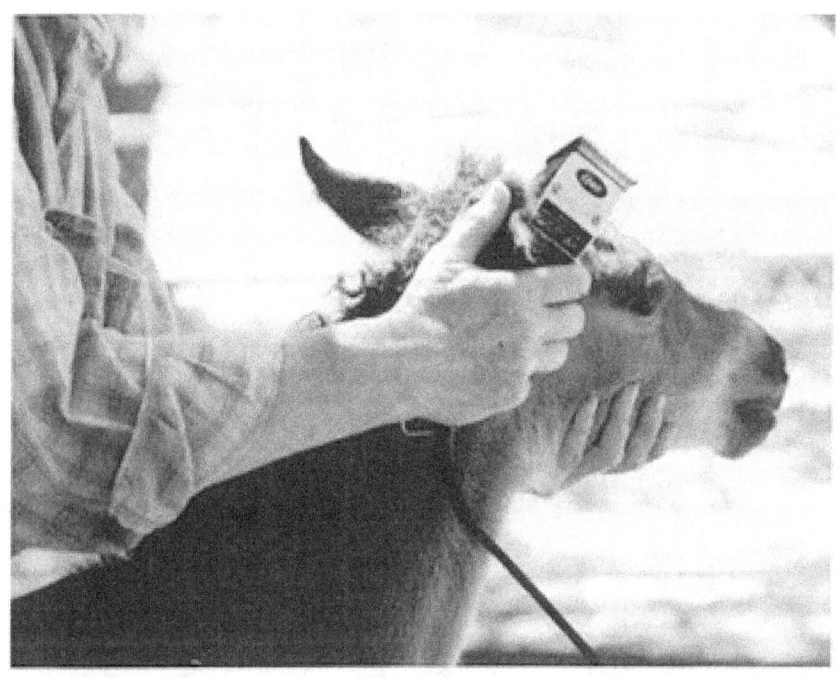

The newborn foal can be taught to tolerate electric clippers, bathing, and fly spray.

Habituating a foal to the noise and vibration of clippers without actually clipping him. When he is older, he will be easy to clip.

HABITUATION

Here, Marty is habituating the foal to the feel of the rope.

But he is careful not to let the foal get tangled in the rope.

Members of the bovine family, whether they be domestic cattle (except for polled breeds), bison, yak, musk oxen, cape buffalo, or water buffalo, have horns. Both the males and the females have them. Their forequarters are massive and powerful to assist in the use of those horns in emergency survival behavior.

The behavior of the very young tells us what survival behavior is. Just as puppies snarl and chew and bite, so do baby kids and calves butt and push and joust, even though their horns have not yet grown in.

Similarly, foals race and run together, because running is the horse's primary survival behavior. Yes, horses can kick, bite, and strike, but their primary instinct when frightened is to run. Look at the horse's anatomy. It is designed to make the horse a running machine, and horses, as a species, are programmed to launch into high-speed flight instantly whenever frightened by an unfamiliar sensory stimulus.

This is the reason that horses run into fences, automobiles, and off cliffs. Their natural habitat is flat, grassy plains. Horses weren't meant to live in box stalls or fenced paddocks, let alone pull wagons or carry riders on their backs.

Why can they be taught to do these things? Because they can be readily habituated to any frightening, but non-painful, sensory stimulus. This can be done at any age, and, in fact, traditionally, most such habituation procedures are done in mature horses. But, it can be done in the young foal in minutes, even in seconds in some cases, and the desensitization to that stimulus is permanent, providing that when it is repeated, it is done in exactly the same manner.

So, when we describe desensitization procedures later in this book, we will see how, in less than an hour, the newborn foal that has never yet been on his feet, can be habituated to saddling, shoeing, bridling, and to having his eyes, ears, nose, mouth, tongue, and feet handled. In that brief time we can prepare the foal for the farrier and the veterinarian so that he will not fear or fight foot trimming, shoeing, the passage of a nasogastric tube (for deworming, etc.), dentistry, rectal or vaginal examination, and handling of the sheath or udder. He will also be taught to tolerate electric clippers, bathing, and fly spray, and all of this can be done before he is one hour of age.

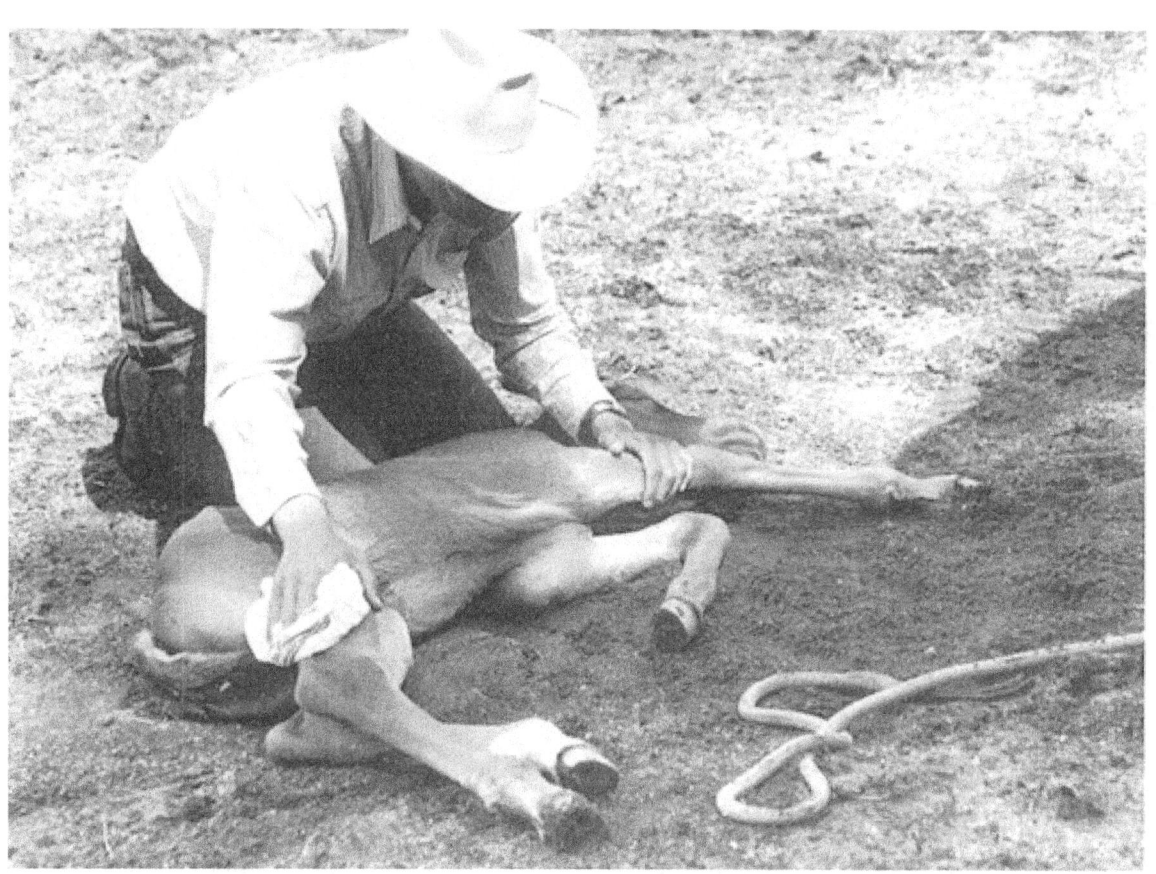

With the foal lying down and totally relaxed, Ma[rk] habituates him to the fee[l] and noise of a plastic ba[g].

We can also get him used to flapping blanke[ts], whirling ropes, [a] water hose, and [a] crackling plasti[c].

Note: No, you do not actually saddle or bridle the foal, or give him a bath. But as a result of desensitizing done now, the foal will later accept those things with little or no resistance.

A little later, when he is steady on his feet, somewhere between 12 and 30 hours of age, we can further desensitize him to pressure in the girth area, so that he will never be cold-backed or cinch-bound. We can also get him used to flapping blankets, whirling ropes, a water hose, and crackling plastic. We can also familiarize him with other kinds of animals if we wish, such as dogs, goats, sheep, pigs, or cattle.

Future police horses can be desensitized to noise, gunfire, and sirens. Military animals can be similarly desensitized to the sounds and sights of battle. Show horses can be habituated to music, flags, and loudspeakers.

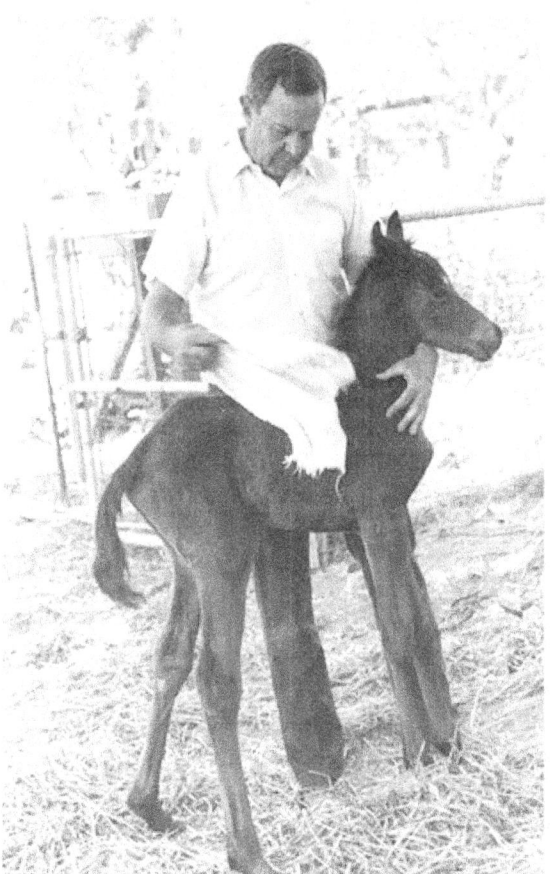

I am habituating this foal to the feel of a cloth rag.

4 SENSITIZATION

All of us have many conditioned responses.

Just as the newborn foal can be habituated to sensory stimuli, he can be sensitized to other stimuli. What do we mean by this? By sensitization, we mean the establishment of a conditioned response.

What is a conditioned response? Remember Pavlov's experiment? Pavlov was one of the first scientific behaviorists. He conditioned dogs to a bell which he rang at feeding time. Initially, the food, not the

The conditioned responses of a cutting horse are developed to a high degree. This is Greg Welch riding Playboys Foxy Gal.

A reining horse must be highly sensitized to the rider's cues in order to spin rapidly, as demonstrated by Bob Loomis, Marietta, Oklahoma.

It is in this manner that a green horse progresses to a world-class horse.

bell, caused the dogs to salivate. Eventually, he could cause them to salivate just by ringing the bell. What had happened? The dogs were now responding automatically. Pavlov had conditioned their sub conscious minds to respond to the bell and cause saliva to flow, even in the absence of food. Even if the dogs had wanted to, their conscious minds could not prevent the salivating because the subconscious part of the brain overrules the conscious part.

All of us have many conditioned responses. In a car for example, even though we are in the passenger seat, we involuntarily stamp on a nonexistent brake if a signal light suddenly flashes red. If we are an experienced driver, we are conditioned to that response, and it is involuntary.

Such actions as driving, writing, and typing involve complex combinations of conditioned responses. Athletes train in order to develop automatic conditioned responses. When you see a good cutting horse work, you are witnessing conditioned responses developed to a very hig degree. A high-level dressage horse and a top reining horse are prime examples of conditioned responses.

A newborn foal can be taught the basics for a turn on the hind quarters, which is required in this trail class maneuver.

This Arabian mare is so highly attuned to her rider that she is stopping on a loose rein. The rider is Tyson Randle, Elizabeth, Colorado.

dressage horse is extremely sensitive to rider cues.

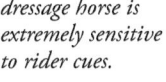

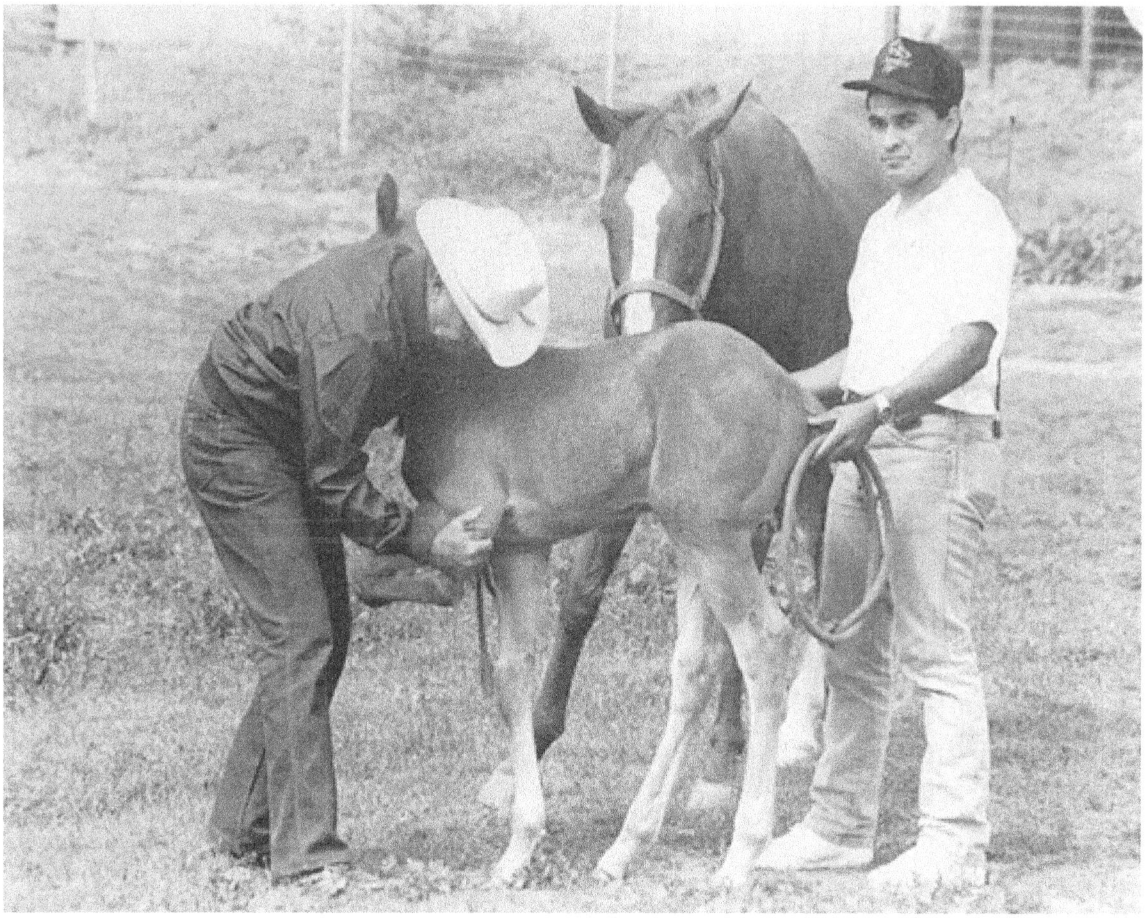

IMPRINT TRAINING

This foal has been conditioned to pick up each foot.

A day-old Welsh foal being conditioned to pick up his feet. I recommend that novices work as a team in imprint training to make sure the foal doesn't escape and learn to be evasive.

Horses can be conditioned to respond this way at any age. What trainers do is cause a stimulus, such as pressure from the bit or spur, until a small response is observed, whereupon they reinforce or reward that response by removing the pressure instantly. The more immediate the reinforcement, the faster the horse will catch on.

When the stimulus is repeated several times, the horse will soon start to automatically respond to the signal. Then, by a process called "successive approximation," the trainer will gradually and progressively increase the response until it reaches the level that the trainer desires, or at least the level that the horse is capable of reaching.

It is in this manner, patiently and consistently, that a green horse progresses from sluggishly responding to the rein to the spectacular spinning of a world-class reining horse. Slow voluntary responses are progressively developed into instantaneous involuntary conditioned responses.

The advantages of developing certain conditioned responses in the newborn foal include the speed with which it can be done at that age, the retention of the learning, and the fact that a responsive, obedient, and easy-to-handle foal is the result.

I do this type of training after the foal is on his feet, usually at one day of age.

There are five conditioned responses that I want, and each can be taught in minutes if properly done. They are:

1. To pick up each foot when asked.
2. To be halter-broke. This means that the foal will follow the halter lead rope without resistance, and that the foal will not pull back against the halter when he is tied. But I do not actually tie the foal. I simply fix the rope so that the foal thinks he is tied.
3. To move the hindquarters laterally when cued.
4. To back up on command, in response to chest pressure.
5. To move forward in response to butt pressure.

IMPRINT TRAINING

A foal should pick up each foot when asked.

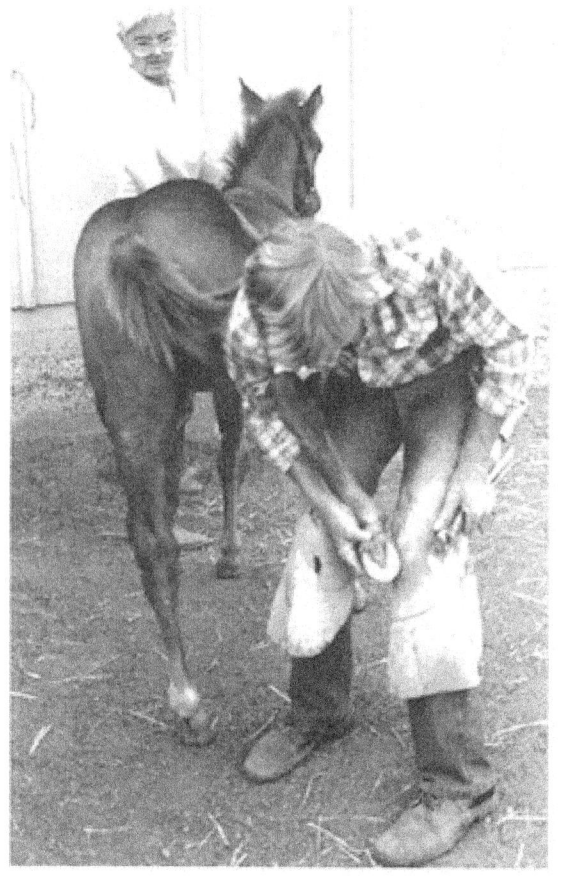

Because this 3-month-old foal was imprint trained, he stands quietly for his first trimming. Rob McCartney, Black Forest, Colo., is the farrier.

SENSITIZATION

This imprint-trained foal has already learned to lead.

I am reinforcing sensitizing procedures on this 3-day-old Quarter Horse/Thoroughbred foal. Here, she has just stepped forward in response to my hand pressure on her hindquarters. As soon as she stepped forward, I released the pressure.

I am teaching this 24-hour-old Appaloosa foal to move his hindquarters to the left in response to pressure in his right flank.

5 DOMINANCE

SOME SPECIES of animals, such as bears and most members of the cat family, are loners. Such animals are together when they mate or when they are immature, but most of the time they are alone.

Other animals naturally live in groups, also called flocks, herds, packs, tribes, coveys, pods, etc.

Human beings are pack creatures, as are baboons and most other primates, plus whales, dolphins, cattle, dogs, wolves, caribou, antelope, horses, and many kinds of birds and other species.

Animals that live in groups have an order of dominance, often called a "pecking order" because of this familiar behavior in domestic chickens. The most dominant individual rules the flock, whereas, at the other end of the spectrum, the most submissive individual is dominated by all the others. This is the dominance hierarchy.

Horses are herd-dwelling creatures and in the wild, the horse herd is led by a dominant leader to whom all of the other horses are submissive. So, it is important to understand that horses are by nature

Animals that live in groups have an order of dominance.

Horses are herd-dwelling creatures.

DOMINANCE

In this particular herd of mares and geldings, two geldings (center) are arguing over who will be dominant.

The bay gelding won, and took charge of the herd.

programmed to be dominated or led. Of course, the naturally submissive horse is easier to dominate, and the superdominant horse is more difficult to dominate, but nevertheless, all horses are capable of being dominated.

It is vital to understand that when we speak of dominating a horse, we do not mean that we must physically dominate the horse. No human is stronger than a horse. Dominance does not refer to physically abusing, hurting, or mistreating the horse.

Horses in the wild state are usually dominated by an old mare. She may even be a decrepit old mare. She dominates the herd by means of the force of her personality, or demeanor. They defer to her because of her experience and her wisdom. Her ability to decide where and when to run may mean survival.

A stallion owns the herd, but he usually runs behind, keeping the stragglers up with the leader. The old mare more often leads the herd.

Being submissive means being respectful, obedient, and dependent. In order to dominate horses, which is essential if they are to perform for us, we do not need to be physically abusive. Instead, we create submissiveness by making the horse feel dependent upon us. The horse that is submissive to the rider is respectful and obedient, not fearful. Fear and respect are not synonymous.

When speaking of dominant human beings, we tend to think of tyrants, dictators, or bullies. But religious and spiritual leaders are also dominant people, yet they may be gentle. Dominance does not require aggressiveness or cruelty. Many entertainers are strongly dominant people, able to sway and influence large numbers of people, but they are not necessarily physically intimidating. Great teachers and statesmen are also dominant.

We must dominate the horse if he is to follow our leadership, to respect us, and to obey our wishes. This doesn't mean that we cannot also feel love or affection for the horse, or that the horse cannot feel love or affection for us. We can love parents, teachers, or leaders, yet we can also have respect for them, obey them, and depend upon them. In short, we can be submissive to them. That is the relationship the human must obtain with the horse.

Horses broke to hobbles become more submissive.

How Do We Dominate the Horse?

Before I explain this, I must say that nothing I have stated thus far is an original thought. All of the preceding information is well established and recognized by behavioral scientists, but now I am going to present an idea that is my own. If it has been previously conceived, I have not heard of it. It is this: When an animal is deprived of its emergency survival behavior, it is forced into a more submissive attitude. I have noted this in a lifetime of working with animals.

Each species has a postural way of signaling submission to members of its own species. The posture for submission is the position of greatest vulnerability. For example, the dog's primary defense is his teeth. The dog's position of defense is to display his teeth in a snarl. He signals

DOMINANCE

This horse, who is hobbled in front and has a hind foot tied up, is totally submissive to the cowboy working on his feet.

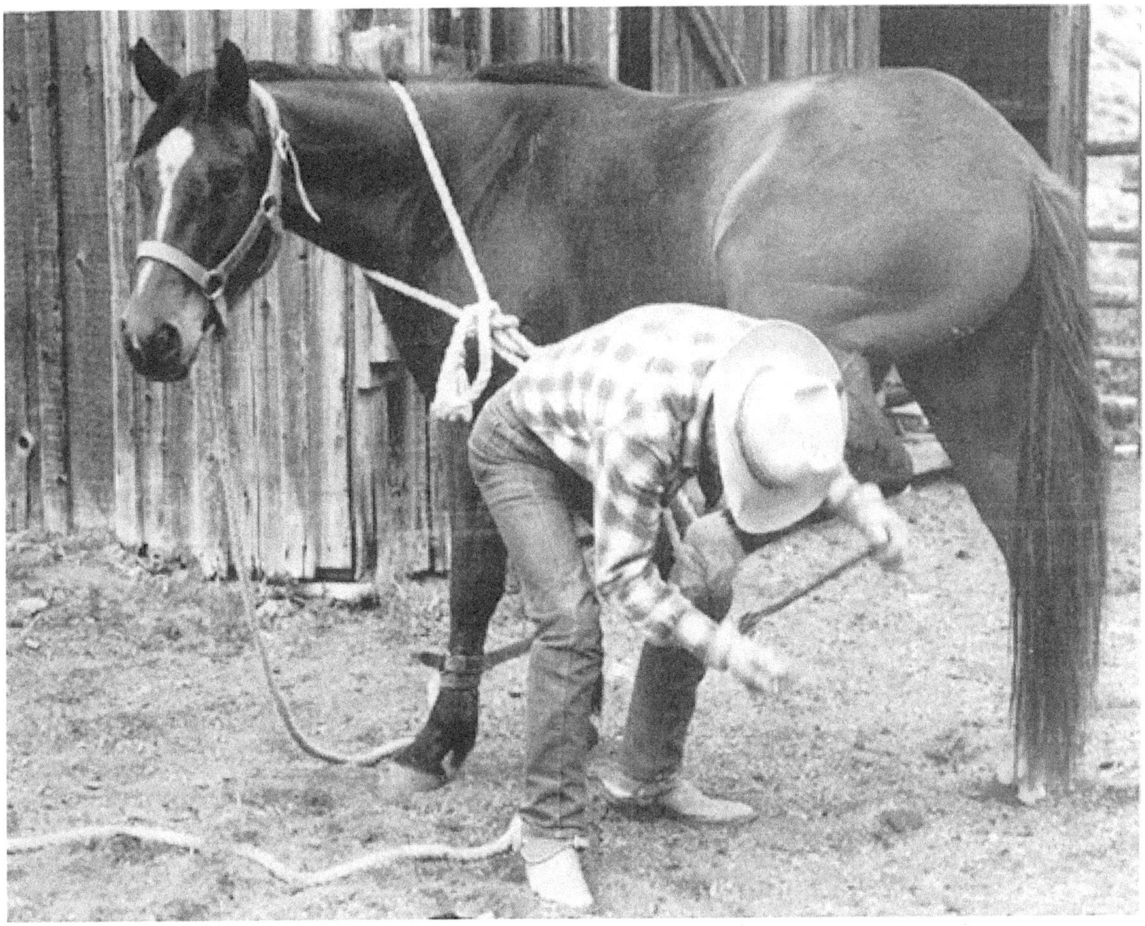

When confined in stocks, a horse usually becomes submissive, providing that he is familiar with stocks.

submissiveness by assuming a position of vulnerability. He lies down and rolls over to expose his throat and abdomen. This behavior is most commonly seen in puppies. If they still feel threatened, they will urinate and finally, as the ultimate act of submission, they will express their anal glands.

Veterinarians know that when a really aggressive dog's mouth is tied shut, depriving that animal of his emergency survival behavior, the dog will then often signal submission by lowering his body position, urinating, and expressing his anal glands.

Cattle use their horns in their primary emergency survival behavior. They signal defense or aggression by lowering their heads and presenting their weapons.

Cattle signal submission by elevating their heads and laying their horns back against their necks. They assume a position, for

IMPRINT TRAINING

Marty Marten is doing a follow-up session with this 18-hour-old, imprint-trained foal, while the mare stands by. Here, the foal is so relaxed he is lying flat on his side.

that species, of vulnerability.

If we immobilize the bovine head in stanchion or in a squeeze chute, or if we elevate the head by means of a nose tong or a bull ring, we deprive that animal of his survival behavior and render him more submissive.

Horses depend upon sprinting away from danger in order to survive. The defensive position for the horse, therefore, is to stand with the legs squared, head up, ready for flight. The position of greatest vulnerability for the wild horse is when feeding or drinking. With his head down, he cannot see, hear, or smell an enemy as readily, nor is he in a flight position. Horses, therefore, signal submissiveness by lowering their head position.

In addition, the horse will signal vulnerability by making feeding motions with its mouth, chewing and licking its lips. This is most obvious in young foals. When they feel threatened by the approach of another horse or a human, they will lower their heads and make smacking movements with their lips. Humans, in order to understand horses, must be able to interpret their

Here, however, he begins struggling to get up, but Marty succeeds in holding him down—thereby creating submissiveness in the foal.

Tying a horse's head to the side not only develops lateral flexion, but also deprives a horse of flight. This creates submissiveness. Trainer Mike Kevil, Scottsdale, Ariz., is working with this filly.

Bitting the horse to develop vertical flexion is another method of flight deprivation.

body language.

Since flight is the horse's primary emergency survival behavior, deprivation of flight will create an attitude of submissiveness in the horse. Anything we do to prevent a horse from running away will make the horse feel dependent upon us. Dependence creates respect. The horse feels vulnerable and needs leadership. If we do not betray the horse by threatening him, he will, when flight is impossible, start to signal submissiveness, enabling us to become a dominant figure.

Although it is not generally recognized, nearly all techniques of horsemanship involve the principle of flight deprivation. We halter break the horse, which, if properly done, conveys to the horse the feeling that flight is impossible when he is restrained by the halter.

We train horses in round pens, working them in circles on a longe line or drive them in one with long lines. We find horses easier to doctor in their own stalls than outside, because they have learned that escape from the stall is impossible.

Horses broke to hobbles become more submissive. Old-time bronc busters tied up a hind foot when saddling a bronc. Brood

Bill Riggins, who was training for the T-Cross Ranch south of Colorado Springs when this picture was taken, shows how he pulls the head around.

mares trained to a palpating chute are usually more submissive in the chute.

Years ago, the renowned "horse tamers" of the 19th century like Rarey, Magner, and Beery all used devices to deprive the horse of flight to make him submissive.

These included casting harnesses, the running W, tying up one front leg, tying the horse's head to his own tail so that he could only circle, and so on. When these flight deprivation techniques were used, "outlaw" and problem horses were quickly subdued.

Sophisticated trainers obtain conditioned responses early in the horses' schooling which will inhibit flight. These include lateral flexion of the head and neck, vertical flexion of the head (head up and nose out is the flight position), and control over the hindquarters. The horse is programmed to automatically move the hindquarters

With the head pulled around, Bill uses his right leg to move the hindquarters to the left. This is a very effective maneuver to prevent a horse from running off—thereby making him submissive to the rider.

laterally in response to leg pressure. When the hindquarters can be moved to the side at the rider's will, full speed forward flight is not possible.

To summarize, deprivation of flight creates submissiveness in the horse. That, in turn, creates dependency, respect, and a desire to follow leadership. When the newborn foal is handled before he ever gets up, the trainer actually is preventing the foal from arising. As a result, right then and there the foal learns to be sub missive to that person and to human beings in general.

Later, when the foal is on his feet and the handler again restrains the foal from running away, the lesson is reinforced, and the foal further learns that the human being is dominant. If, meanwhile, every thing is done with kindness, and the foal is never injured, we have the beginning of the perfect relationship of horse to human submissiveness without fear, plus trust, dependency, affection, respect, and a desire to follow that person's leadership.

In any group of horses, there is always a pecking order. The dominant ones always drink first, while the more submissive ones patiently wait their turn.

These two are arguing over who's the boss, and, therefore, who gets the hay.

Photo by Nancy Clifton

6 THE MARE

WHETHER OR not imprint training of the foal is going to be done, the broodmare should be gentle, tractable, and well halter-broke. She should allow us to handle any part of her body. It is not unusual to find broodmares that are not broke to ride. Many are barely halter-broke, and some cannot be handled at all. The owners of such mares will often excuse their negligence by explaining that "she is just a pasture mare" or "a range mare" or "just a broodmare"

If a broodmare is so unschooled that she cannot be handled, it indicates one of three things:
1. The owner is too lazy to train the mare.
2. The owner is incompetent to train the mare.
3. The mare has a naturally refractory disposition.

If the first or second categories are the case, the owner should hire a competent trainer to teach the mare to lead and to allow handling of her entire body. This

The broodmare should be gentle, tractable, and well halter-broke.

Broodmares and foals in pasture

THE MARE

At foaling time, many mares are put into a foaling stall where they can be observed. The stall should be roomy enough, however, so the mare is not forced to lie next to a wall—which makes the foaling process difficult. This stall is too small.

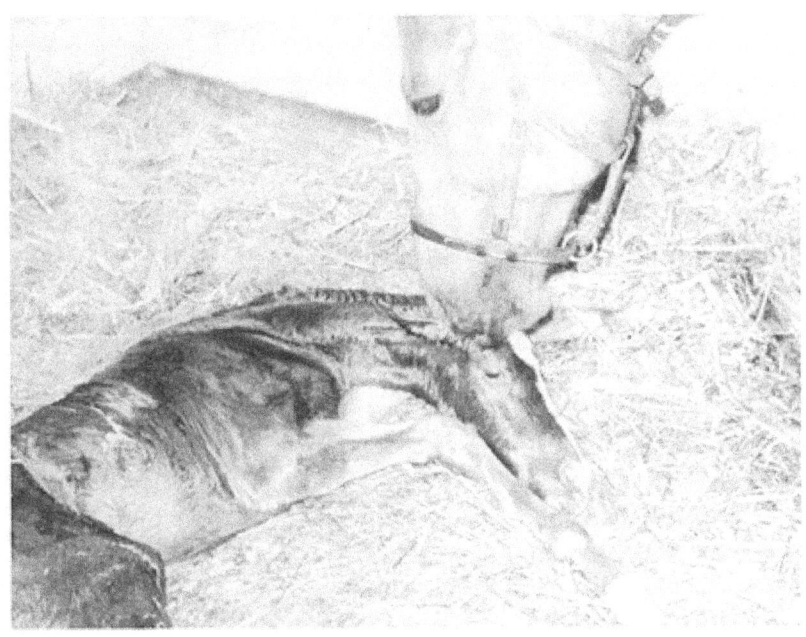

A mare bonding with her just-born foal.

can usually be done in a single lesson if the trainer is truly competent and uses proper techniques.

If the mare's basic temperament is at fault, then she should be disqualified as a broodmare, because temperament is highly inheritable, and because foals wil learn to mimic the behavior of their dams to a great extent. There are, admittedly, exceptions to disqualifying such mares when they are possessed of extraordinary performance talents such as racing speed.

The ill-mannered broodmare will not only tend to adversely influence her foal's behavior, but will also cause a host of other problems. Such mares are hard to load into a trailer or van, cause trouble during transport, and are a problem at the breeding farm. They are difficult to examine for fertility and pregnancy, and difficult to treat if treatment is necessary. Unruly mares are much more likely to injure themselves when being handled, and often injure people who must handle them. If they are unfortunate enough to suffer obstetrical difficulties or develop postpartum complications, they can be very difficult for the veterinarian to treat.

IMPRINT TRAINING

This cranky mare is reluctant to let Marty Marten approach her partly because she is being protective of her 18-hour-old foal.

But the oats win her over—and she comes up to Marty.

THE MARE

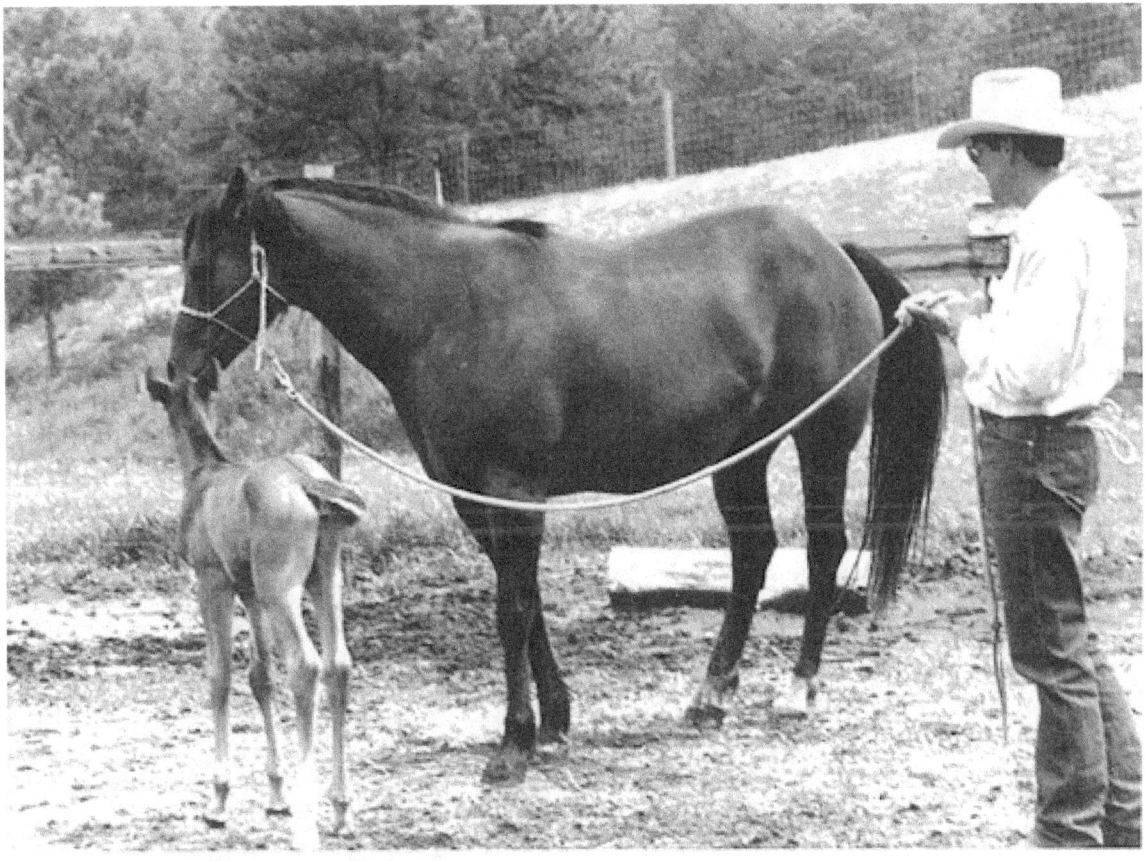

The mare allows Marty to handle her but as soon as the foal moves away, she becomes agitated. This mare was extremely protective of her foal for several weeks. But by weaning time, she was happy to bid goodbye to the little guy.

The foal of an unschooled mare is not a good candidate for imprint training. Her attitude will interfere with the training process and will to some extent negate the efforts expended upon the foal.

Broodmares should be gentled, trained, and taught good manners before they foal, preferably before they are bred. The mare owner who takes pride in how difficult she is to handle is a poor horseman. Moreover, the proven performance mare is the one more likely to pass on her ability to her foal. If a mare has not proven her performance ability, we can only hope and assume that her conformation and bloodlines will give us the kind of foal we want.

When the mare is ready to foal (how that is determined is beyond the scope of this book), she should be brought into the stable. While it is true that most mares can foal in pasture and produce a healthy foal, for our purposes we want the mare to foal where we can observe her. It is an advantage to have running water and electricity available. If obstetrical complications arise, they can be seen and assistance given.

Additionally, the training of the foal can begin as soon as he is born.

7 Immediate Postpartum Procedure

Photos by Debby Miller

WHEN THE foal is born, allow the umbilical cord to break of its own accord. Usually this occurs when the mare gets back to her feet, and this event should not be hurried. The stump of the cord should be disinfected. I recommend soaking it in a "tamed" povidol iodine solution. A towel may be used to rub the foal dry, and to remove obstructing membranes or mucus from his mouth and nose. Halter the mare; ideally, one person should be assigned to control her. Allow her to lick and smell the foal, even as he is being handled and trained. Be sure that the mare's head and the foal's head have proximity. This reduces the anxiety of the mare. Try to keep mare and foal head to head and avoid placing yourself between them.

Now we are ready to begin the training procedures. While we are doing this, the foal may attempt to get up. Do not allow him to get up. Gently, but firmly, keep him on his side. As explained previously, this alone will create an attitude of submissiveness in the foal. Start with the head. Rub the entire face and head. If the foal resists, do *not* stop. Persist in your manipulations until the foal relaxes. This indicates that habituation has occurred. Continue the rubbing beyond this point, however.

This Is Very Important: You cannot overdo the stimuli, but you can underdo them. If you stop the stimuli while the foal is trying to escape, you will fix that behavior. In other words, you will sensitize the foal to

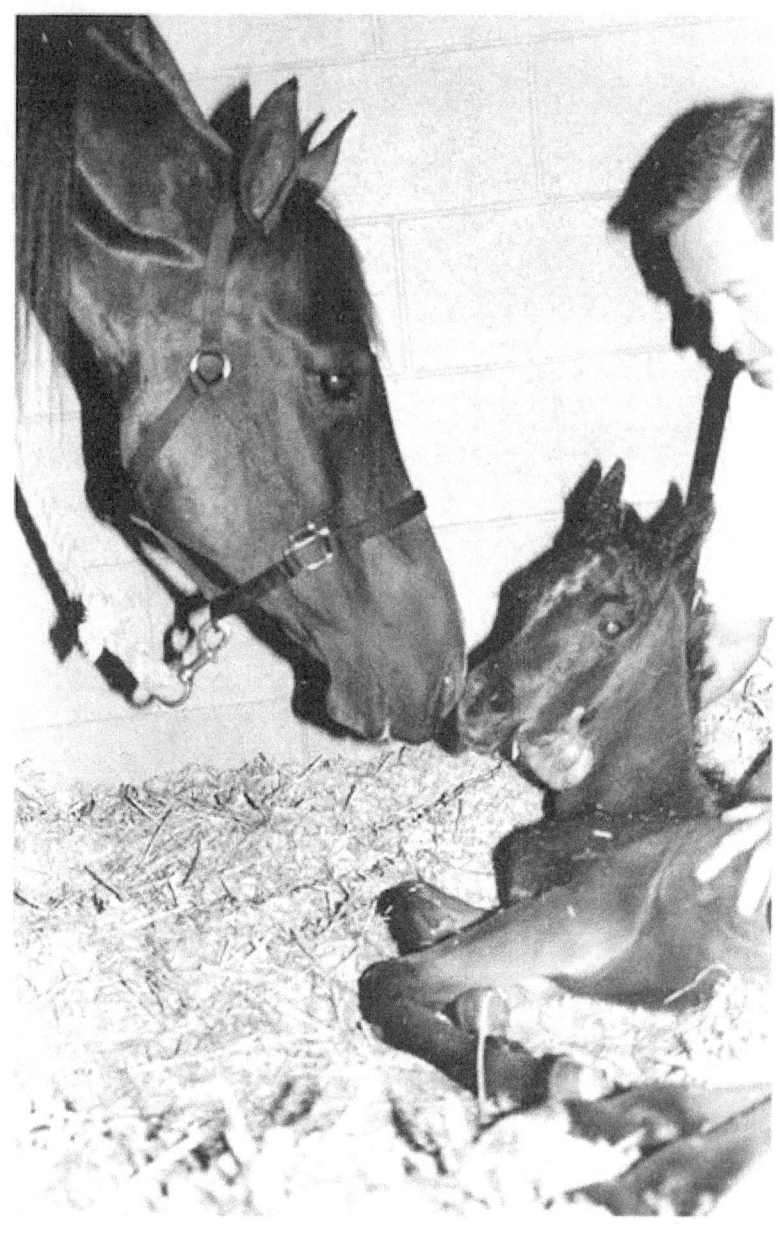

This is a newborn Arabian filly on Ventura Farms, owned by David Murdock and located in Thousand Oaks, California. The mare and foal are bonding in this picture.

IMMEDIATE POSTPARTUM PROCEDURE

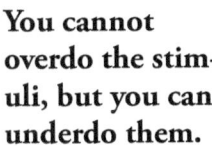

I immediately begin imprinting the foal, starting by desensitizing the face.

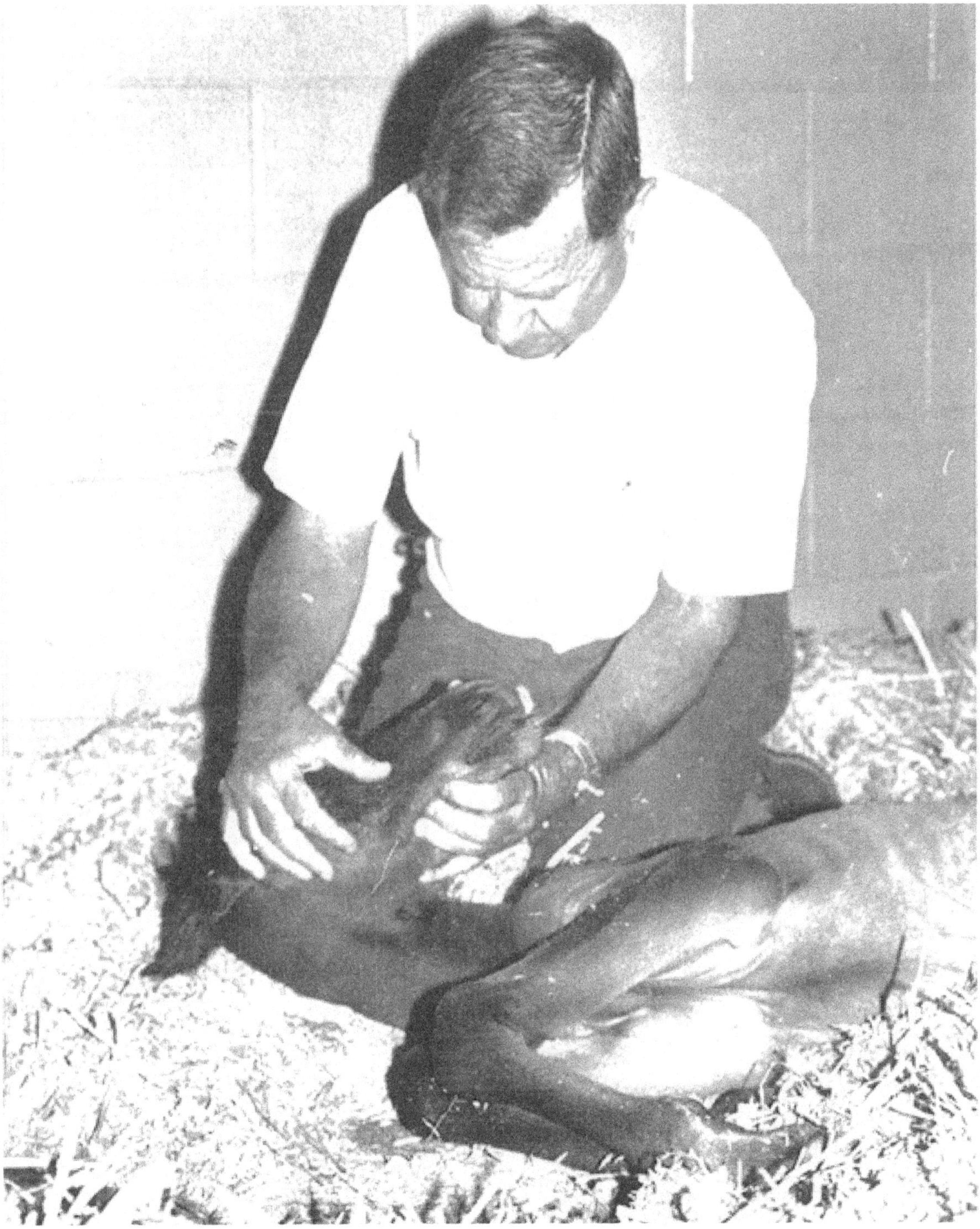

You cannot overdo the stimuli, but you can underdo them.

that stimulus, and right now we are trying to desensitize or habituate him.

Once again, each stimulus must be repeated until the foal no longer resists, and he relaxes and becomes apparently oblivious to it. Do not rush imprint training. If the handler becomes fatigued or must leave for some reason, stop the training routine. It can be resumed later, but do not rush any specific procedure. Persist until after habituation occurs. This will usually require from 30 to 100 repetitions.

After desensitizing the face and head, including the poll, do the ears. Rub and massage each ear until the foal is desensitized to having his ears touched. Then insert a finger into each ear canal, and wiggle it to desensitize that area.

43

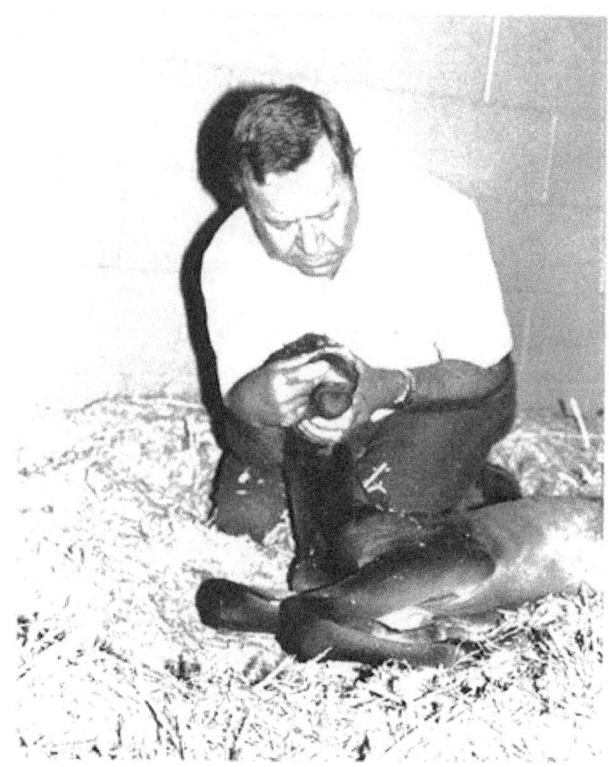

... the right nostril ...

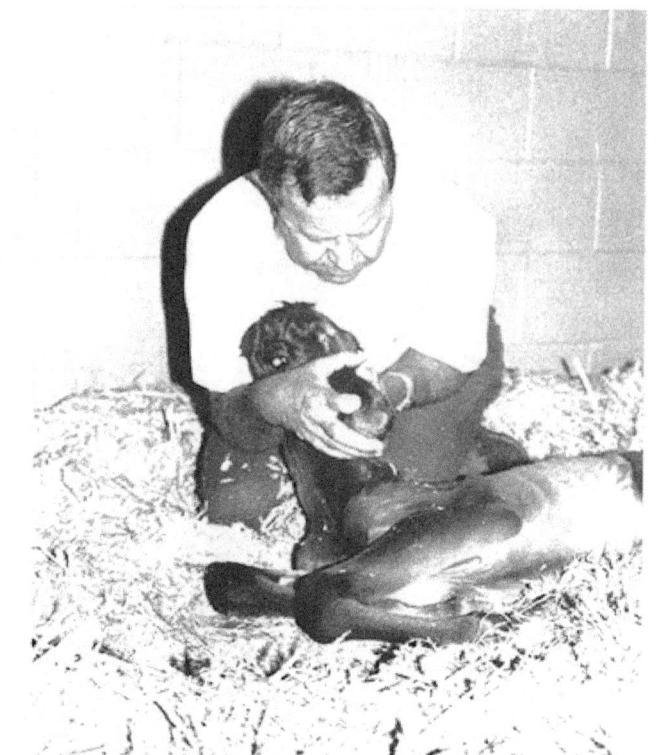

... the underside of the lip ...

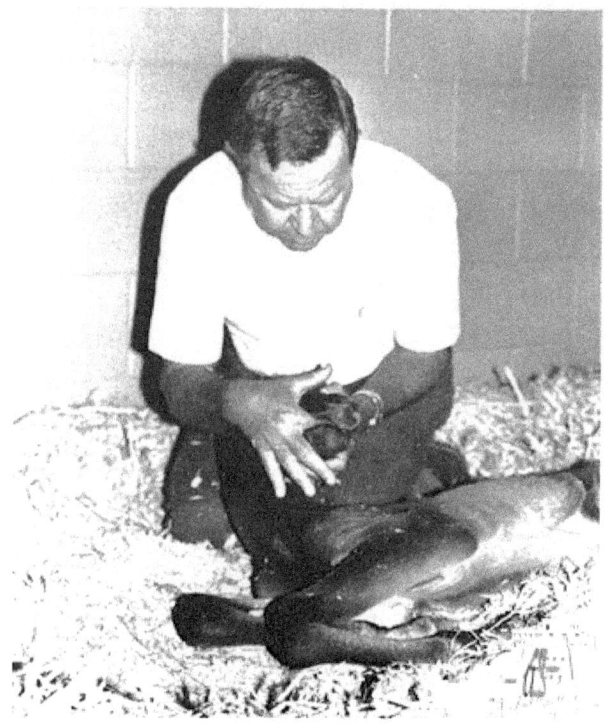

... the mouth and tongue ...

IMMEDIATE POSTPARTUM PROCEDURE

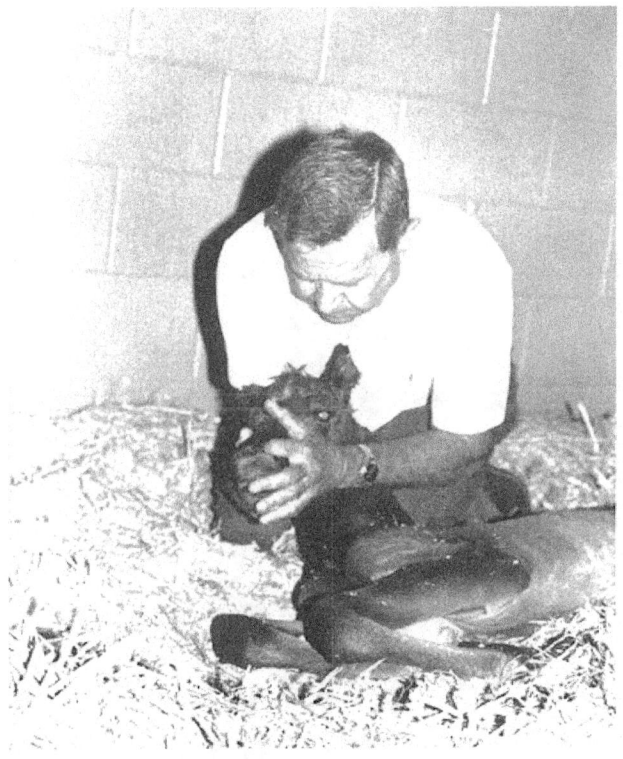

. . . the left nostril . . .

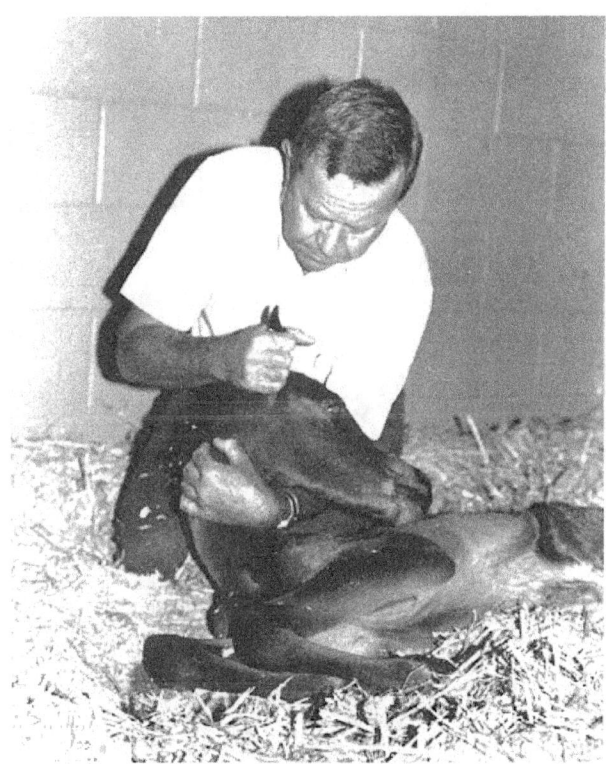

. . . the external ear . . .

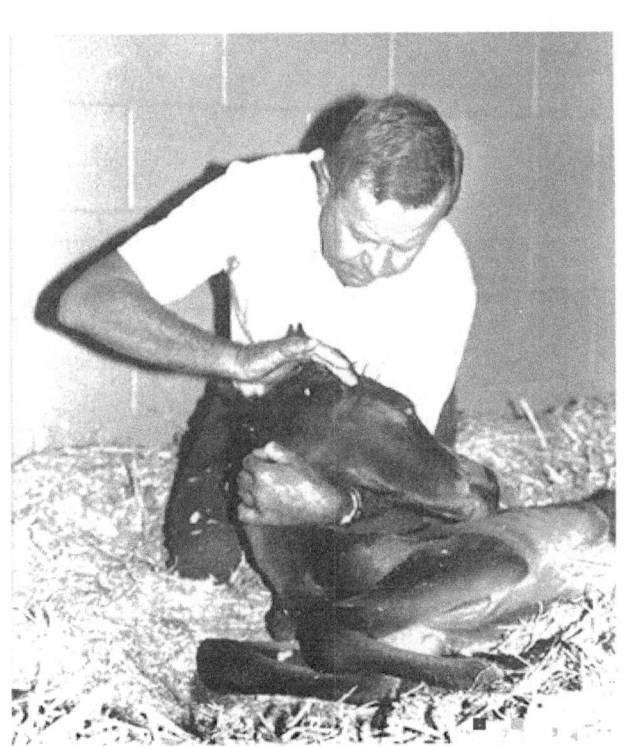

. . . and the inner ear. Note the position of the head in this and the previous photo. There are two important reasons for this position: First, it keeps the foal from getting up, thus establishing my dominance.
Second, it conditions the foal for lateral flexion, which will facilitate teaching the filly to lead in the second and third training sessions later on.

IMPRINT TRAINING

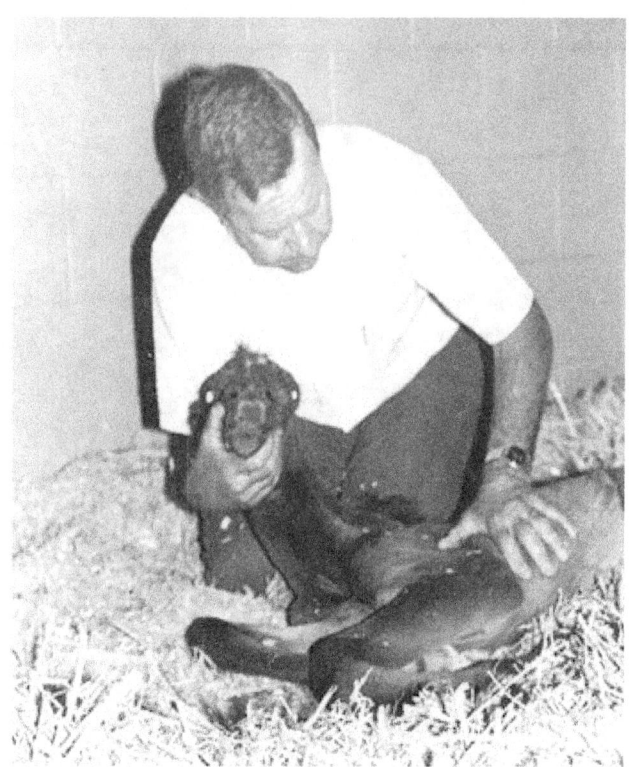

... Desensitizing the left shoulder and chest wall ...

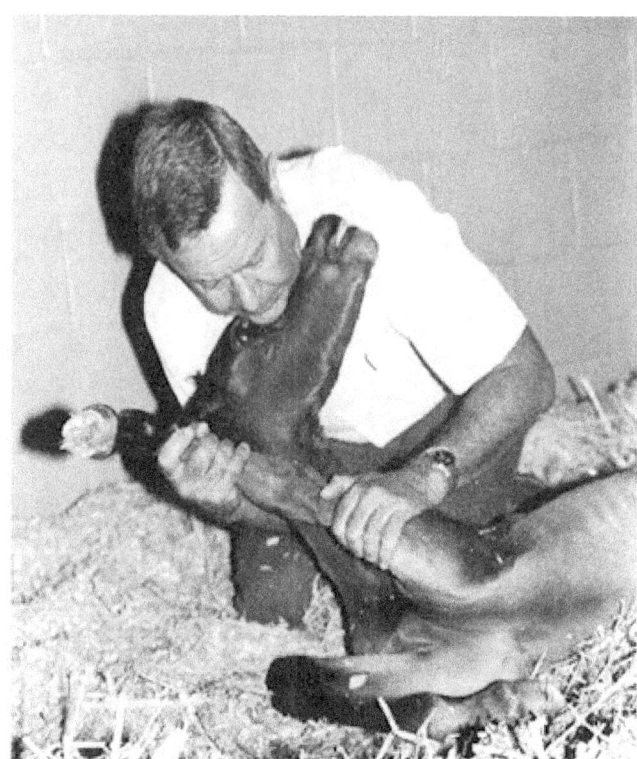

... and the left foreleg. Note how I restrain the head with my chin. Ideally, it helps to have an assistant, especially on big, strong foals.

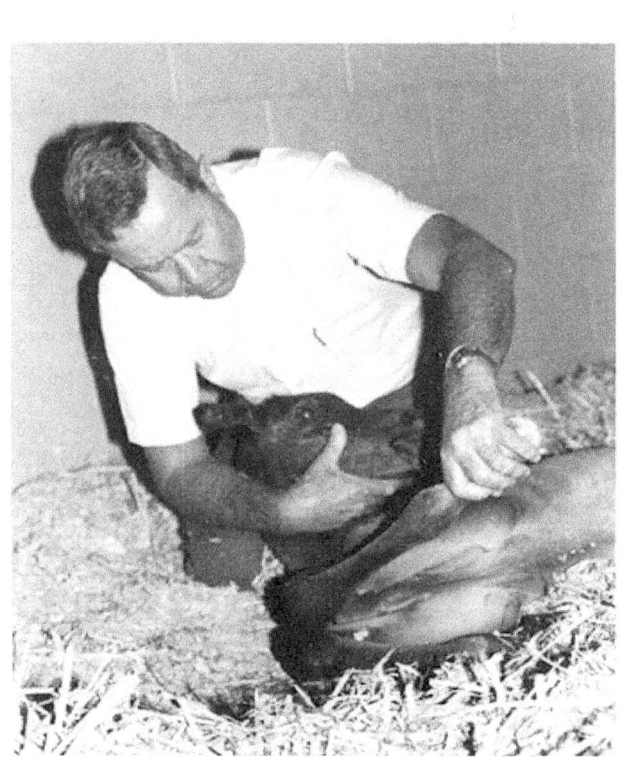

Repeatedly flexing the left fore leg until it relaxes

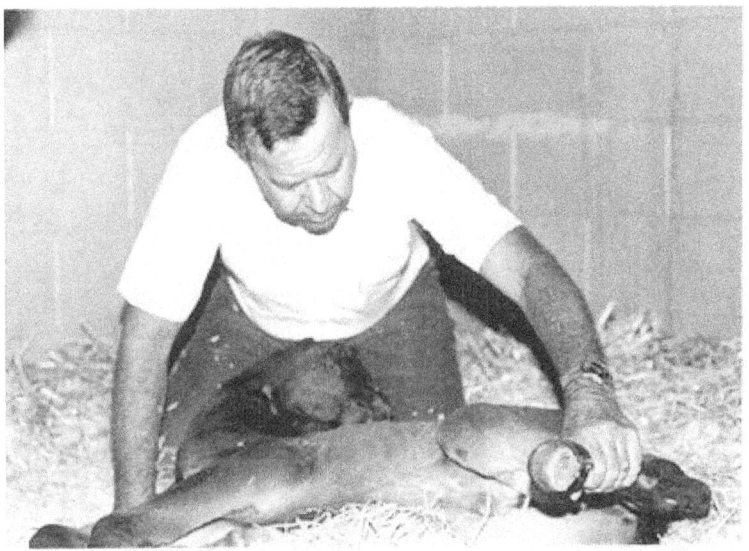
Repeatedly flexing the left hind leg until it is relaxed and yeilds freely.

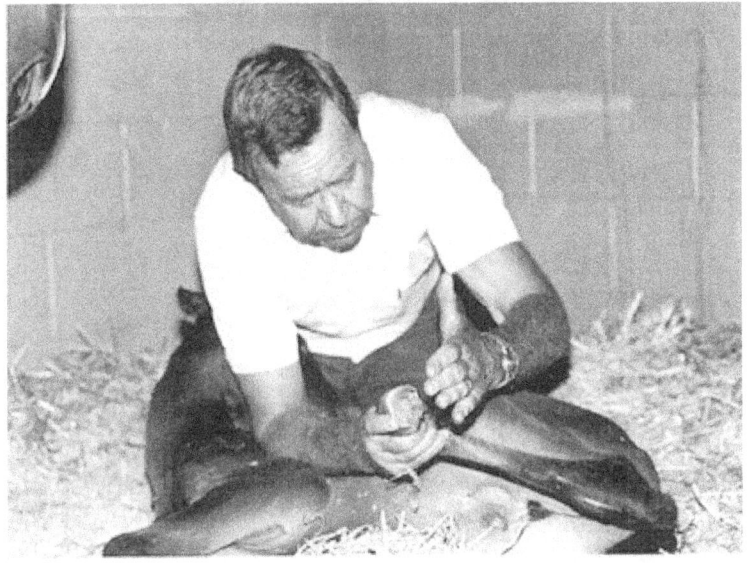
Tapping the left hind foot 50 times to desensitize the foal to having that foot trimmed and shod

Desensitizing the left forefoot by tapping it 50 times. The filly is trying to get up, but can't because of the way I am restraining her.

Next, do the nostrils. Gently insert a finger up each nostril and rhythmically wiggle it until it is desensitized.

The mouth will be next. Massage the upper lip. Do the underside of the upper lip. Then insert a finger into the mouth and desensitize the mouth and tongue. Your foal will always be easy to put a twitch on, bridle, or perform dentistry upon. While you are doing the mouth, the foal will probably make nursing movements with his mouth and tongue. Ignore these.

Having completed the head, we proceed to the neck. Do all sides of the neck, including the mane. Take plenty of time. It will take a full hour to properly desensitize the newborn foal that has not yet been on his feet.

After the neck has been done, desensitize the withers and the back, all the way back to the base of the tail. Then do the tail and the perineum, the area under the tail.

Because the foal is lying on his side, do the upper shoulder, rib cage, and the chest. I do not try to desensitize that part of the abdomen where, eventually, the rider's heel and/or spur will touch, because after the foal is on his feet, we will *sensitize* that area.

Desensitizing inside the back legs and the udder area.

Desensitizing the tail . . .

. . . and the perineal area.

Desensitize the groin area.

Testing the left hind leg to see if habituation is adequate.

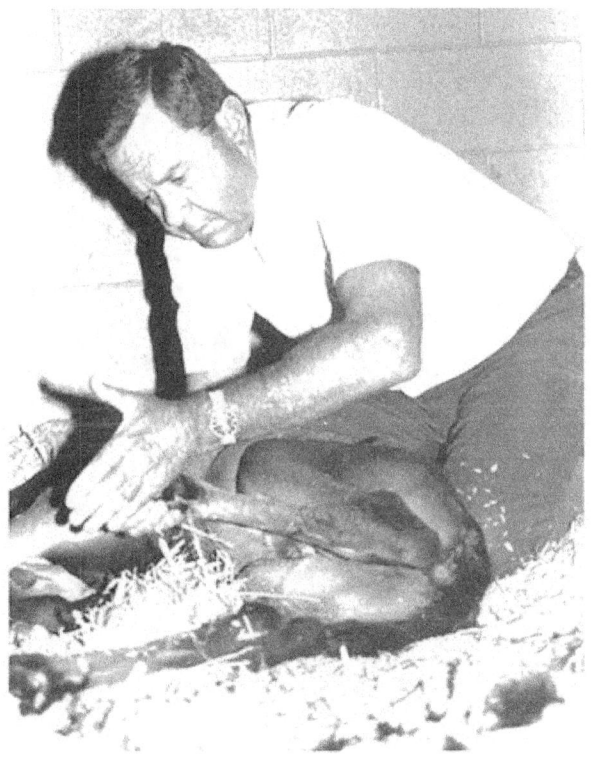
Testing the left hind foot.

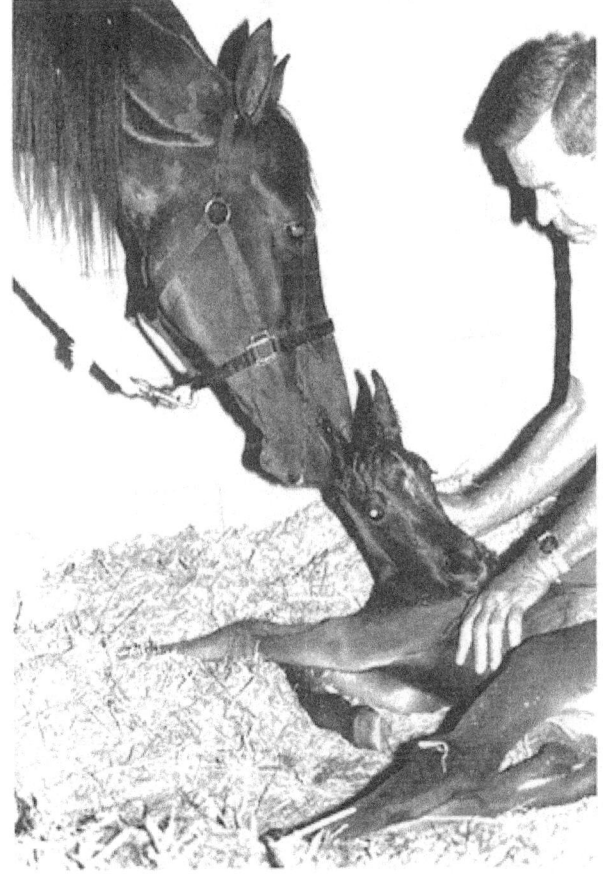
Pausing to let the mare and foal bond a little more

IMPRINT TRAINING

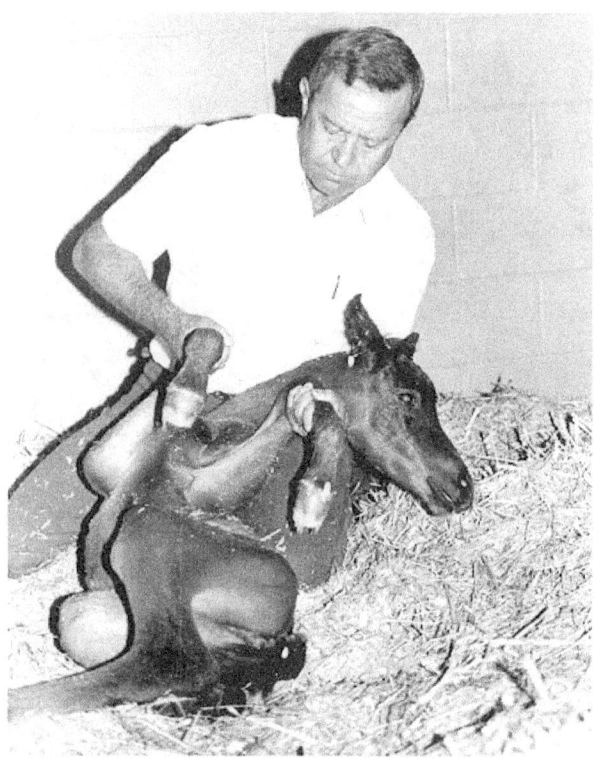

Turning the foal over.

I am now ready to repeat the entire process on the other side.

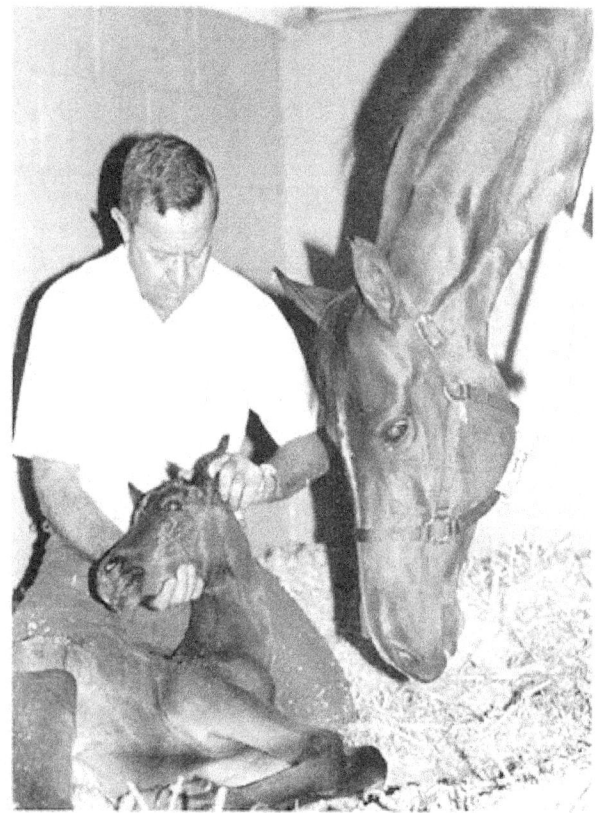

Desensitizing the left ear canal by inserting my finger into it 50 times.

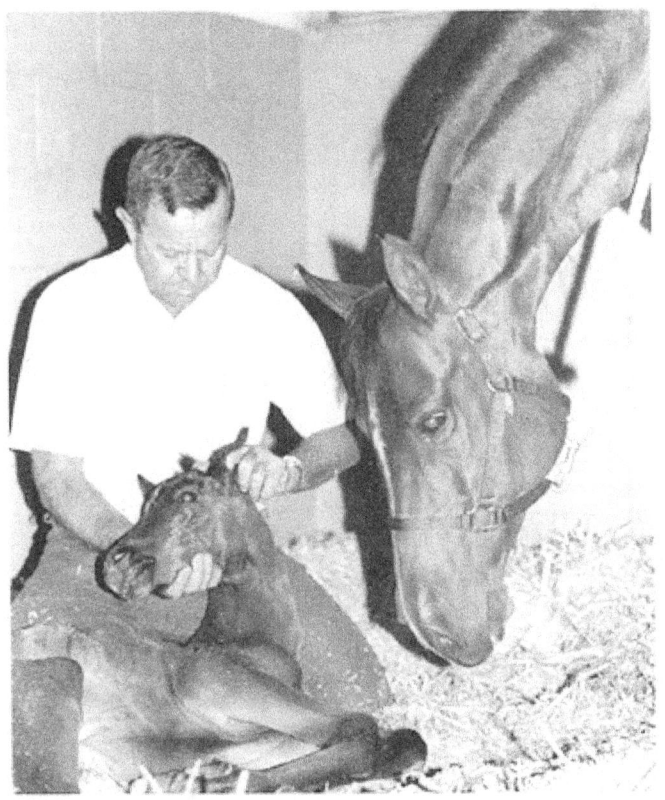
Desensitizing the left external ear. Note how the head and neck are flexed to the right in these photos

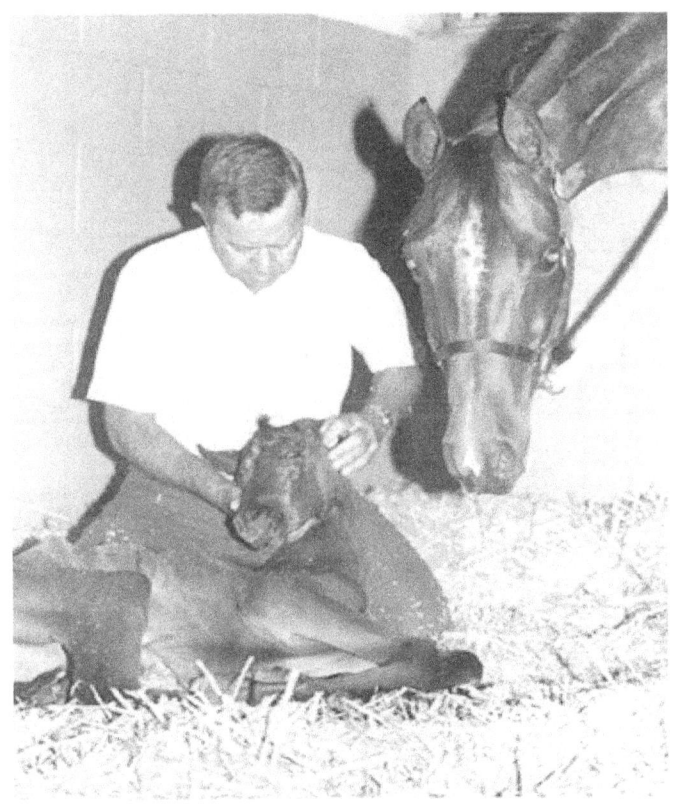
Desensitizing the right front foot by tapping it 50 times.

Do the uppermost limbs, stroking their full length repeatedly until they are completely desensitized. Then repetitiously flex each joint (elbow, hock, stifle, etc.) about 30 times. Finally, and this is very important, pat and rub the bottom of each foot at least 50 times. This teaches the foal not to fear shoeing and rasping. Pat the bottom of the feet hard enough to make an audible slapping sound. You can even tap on them gently with a small metal tool to simulate hammering.

Desensitize the groin area. If the foal is a male, be especially thorough about the sheath. If it is a filly, pay attention to the udder region.

Without allowing the foal to get up, roll him over and repeat the entire procedure on the opposite side. Watch out that his flailing legs don't kick you.

It is also a good idea to desensitize him to electric clippers. Do not clip the foal, but, with the motor running, rub the foal's entire body with the clippers. Be especially thorough around the face and ears. I like to use small clippers first, and then switch to big, powerful, noisy clippers. As always, persist in each stimulus until well after desensitization occurs.

If desired, you can use a spray bottle with warm water to desensitize the foal to being sprayed with fly repellent. Use discretion, however. Spray the foal very lightly—do not get him wet. Or, if the weather is damp and chilly, it might be wise to postpone this step.

I also like to rub the foal down with a piece of crackling, white plastic. White is the most visible color to a horse, and the crackling of plastic, if you think about it, must sound like a lion going through dry grass. No wonder so many horses are afraid of paper and plastic, and get spooky on a windy day.

IMPRINT TRAINING

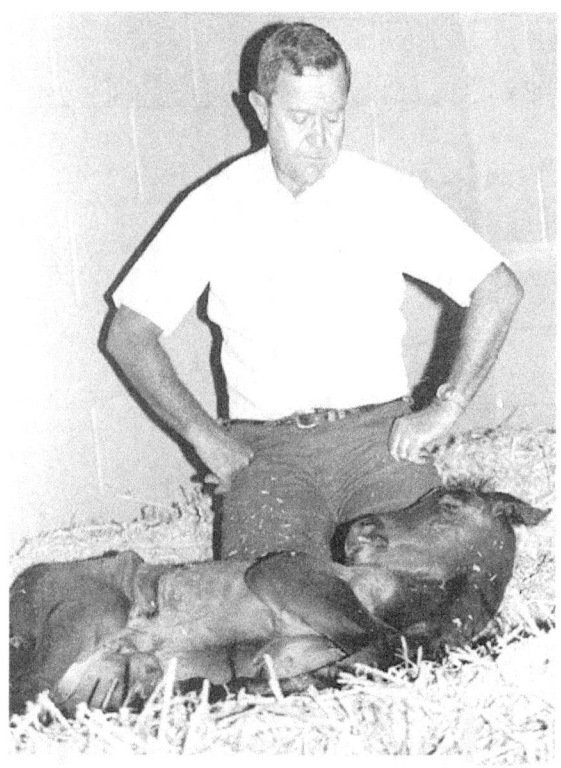

Note how the filly lies quietly with her head turned to the right. She no longer struggles—and is relaxed, calm, and submissive. I have now completed desensitizing all of the areas on the right side of the filly.

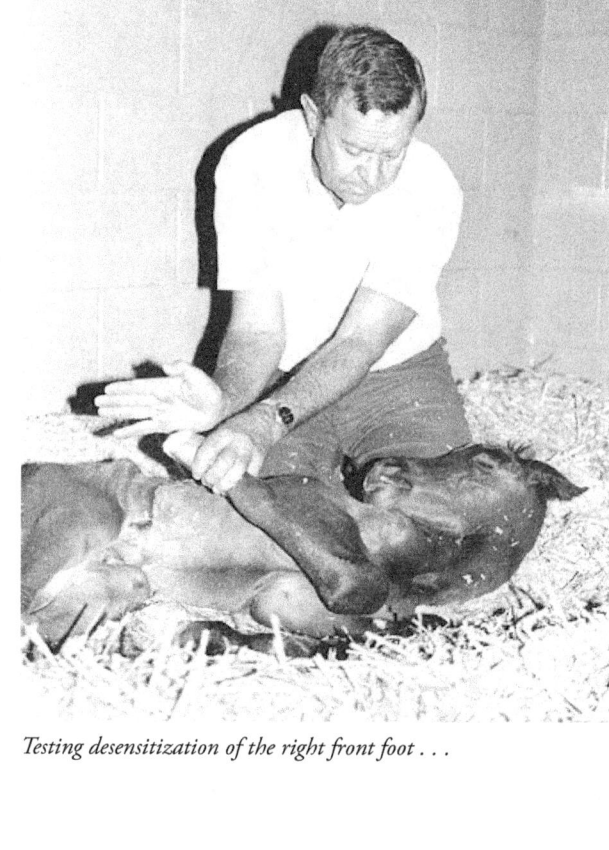

Testing desensitization of the right front foot . . .

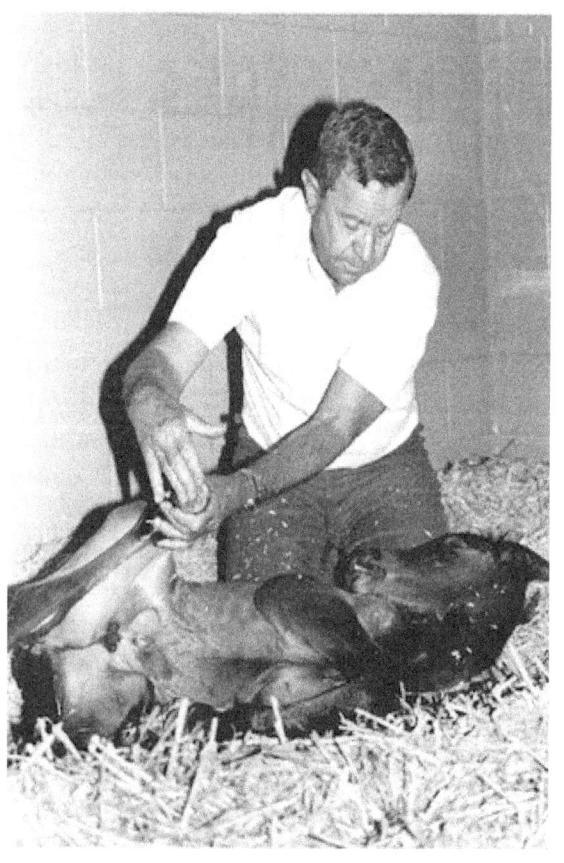

. . . and the right hind foot.

All of this will have taken approximately an hour. While we are doing it, allow the mare to smell and lick the foal. We are nearly finished with our initial imprint training at this point. Before we leave, however, allow the foal to stand and acquire strength and balance, and to nurse.

IMMEDIATE POSTPARTUM PROCEDURE

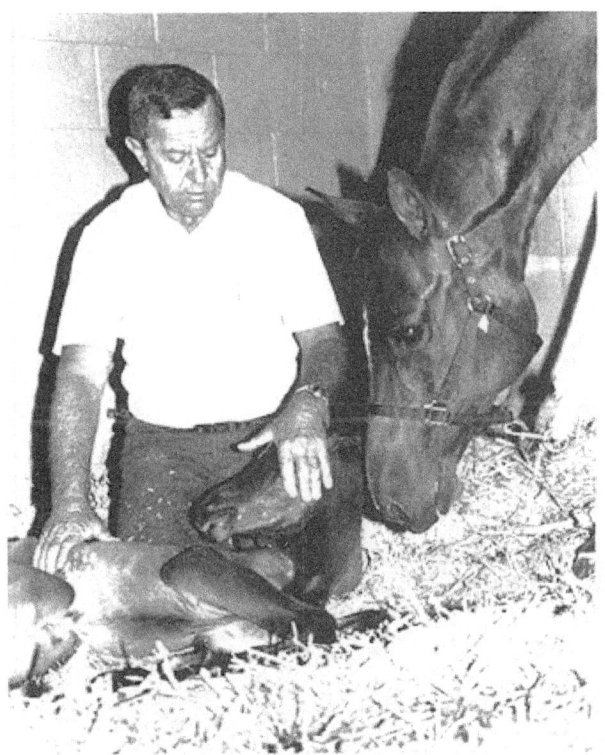

Allowing the mare time to bond with the filly.

Desensitizing the filly to a flapping white cloth . . .

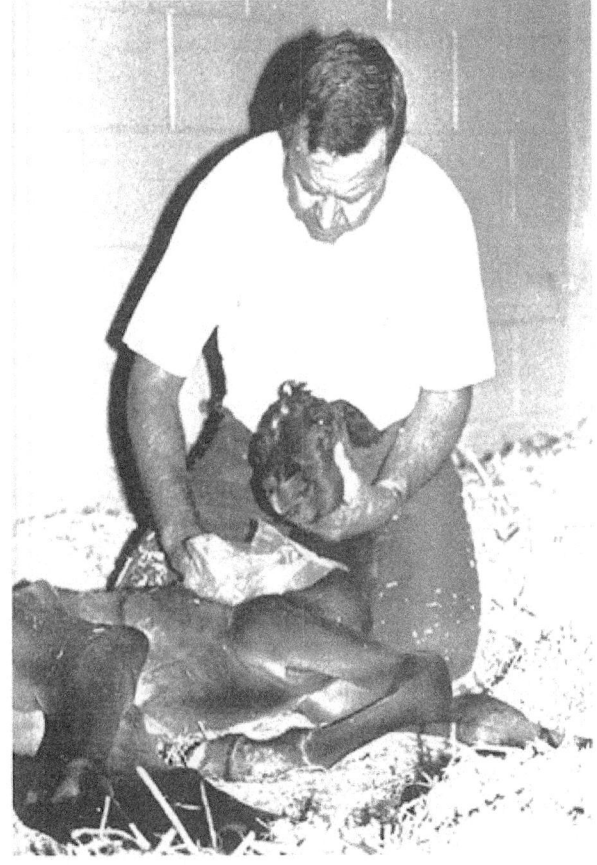

. . . and a cracking plastic bag on her body . . .

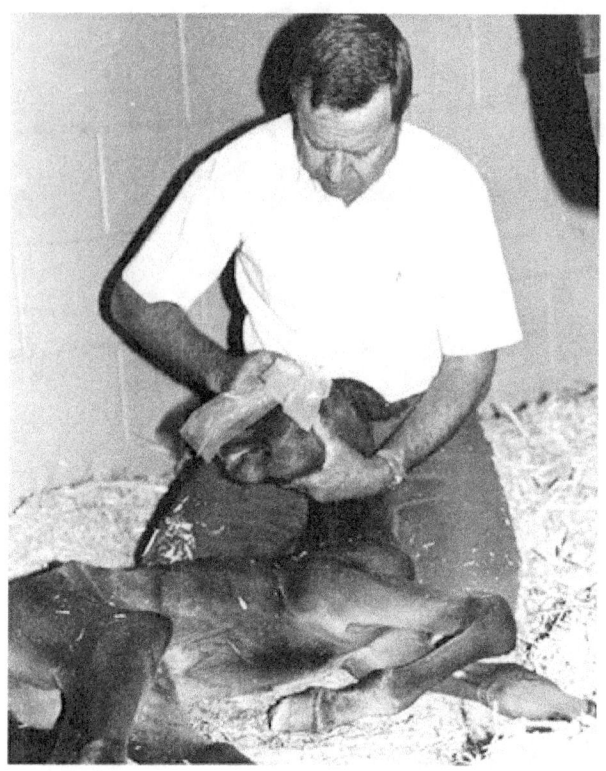

... and on her face.

Desensitizing the foal to the hum and vibration of electric clippers along the side of her head ...

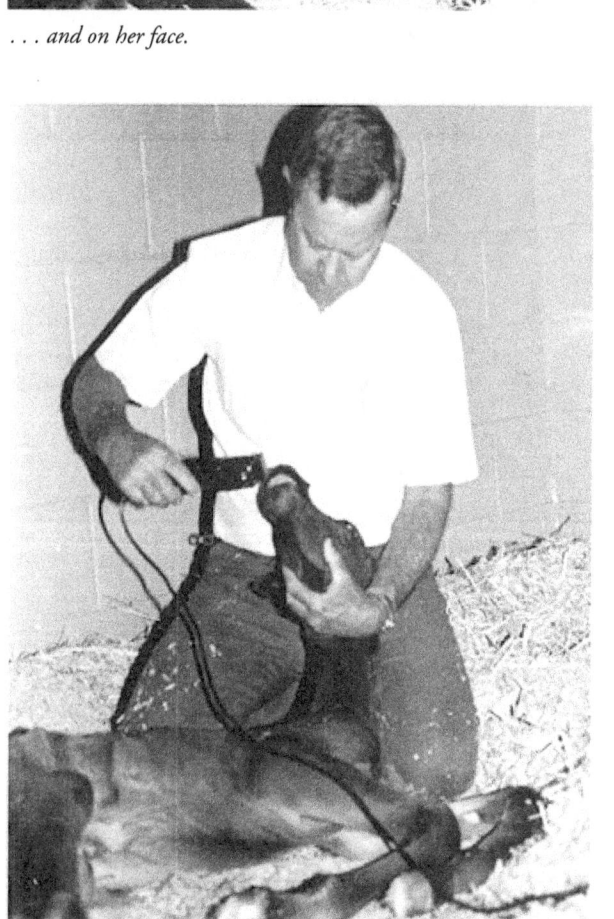

... around the lips and muzzle ...

IMMEDIATE POSTPARTUM PROCEDURE

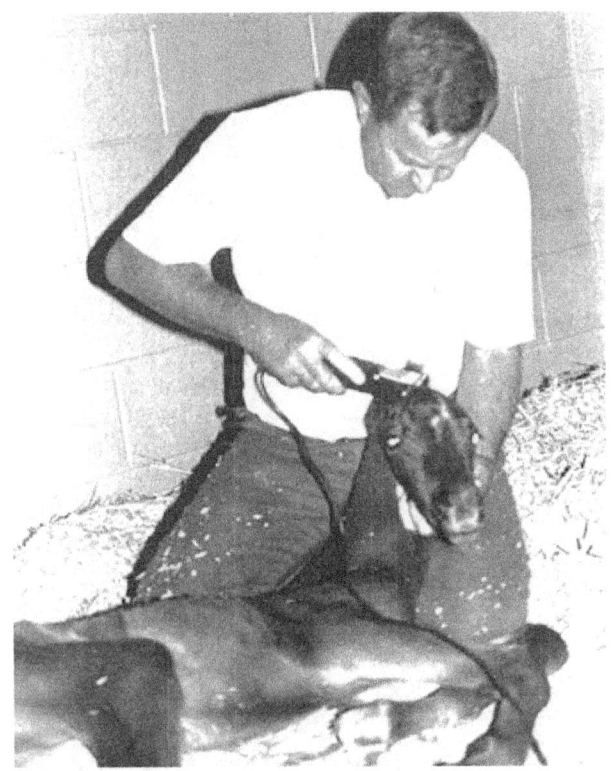

... the right ear ...

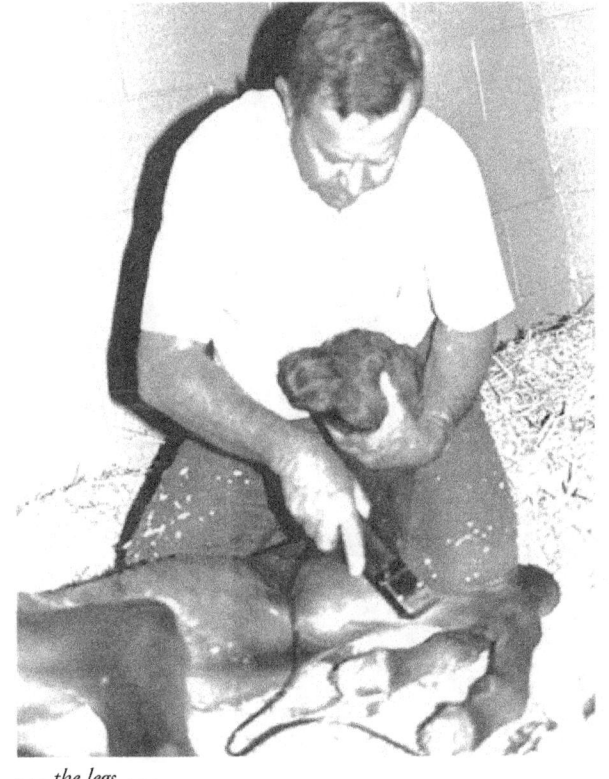

... the legs ...

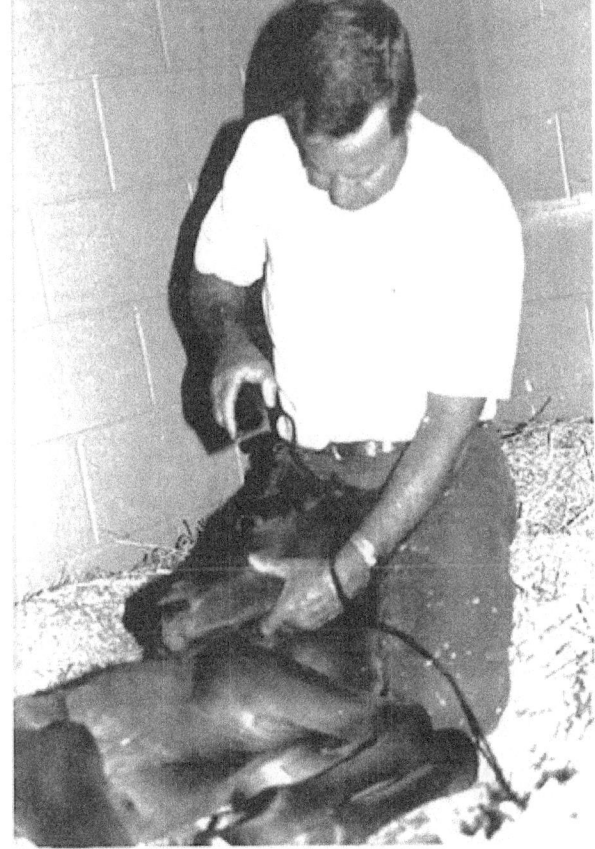

... and the left ear ...

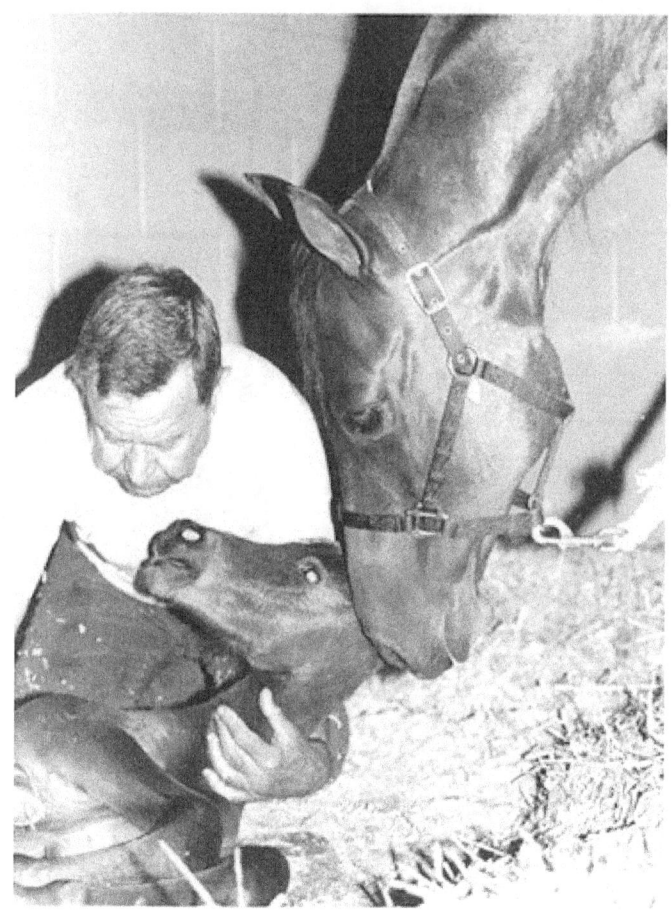
The mare continues to bond with the foal as I work with her.

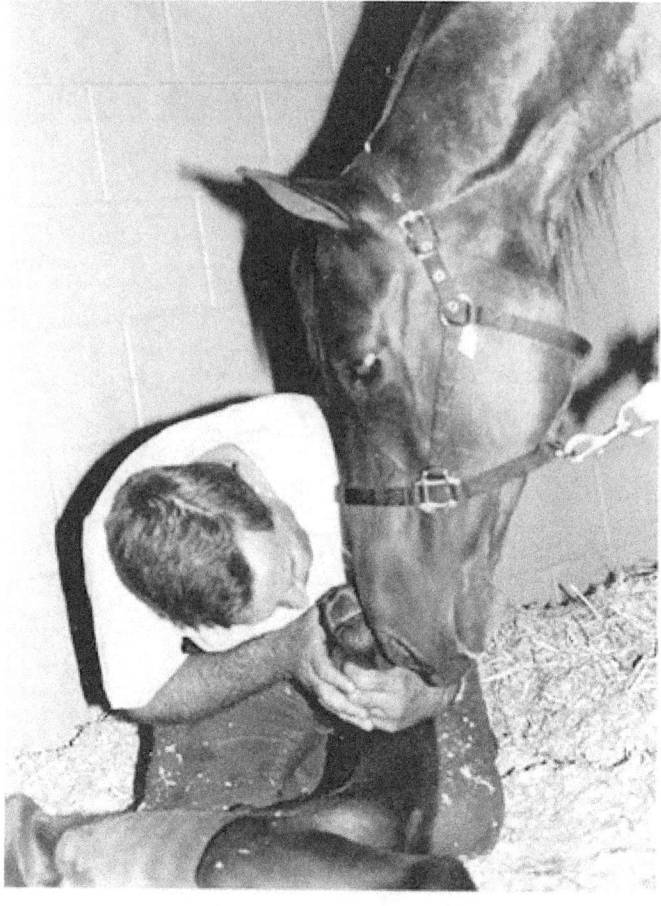
I blow into the filly's nostrils to facilitate her bonding with me

Suppling the Neck and Legs

When working with the newborn foal, as he lays on his side, I bend the head to the side, bringing the nose toward the withers. I usually maintain it in that position by tucking the nose under my arm, using my elbow to keep it there for a few minutes while my other hand rubs the foal and does the various tactile flooding procedures. Since I roll the foal over in order to equally desensitize both sides of his body, this lateral flexing of the head and neck is done in both directions.

My original reason for doing this was simply to prevent the foal from getting up. I usually do the imprint training by my-

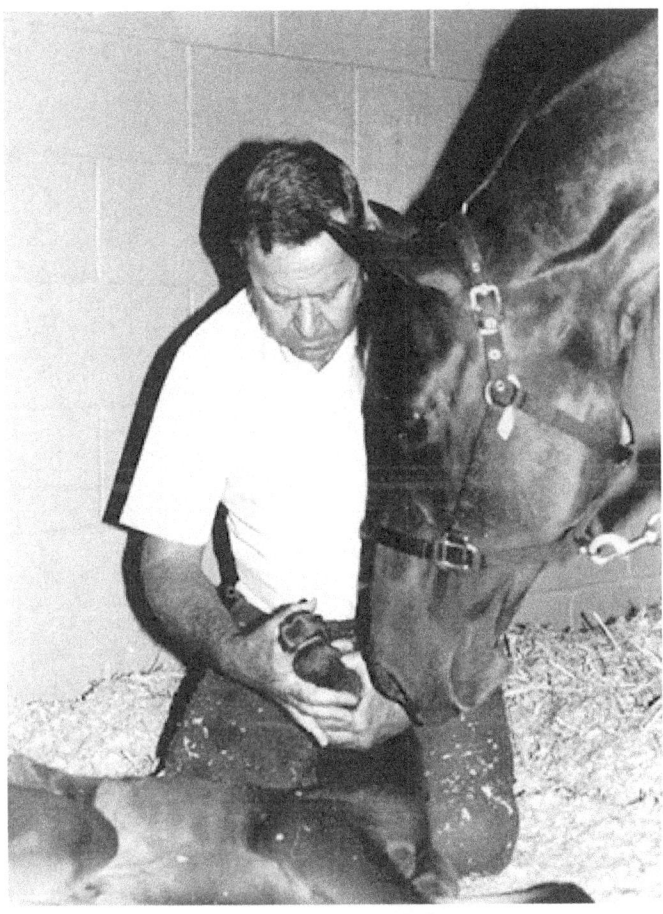
Desensitizing the mouth again—from the right side this time. From now on, it will be easy to examine the mouth of this filly. She will also be easy to bit.

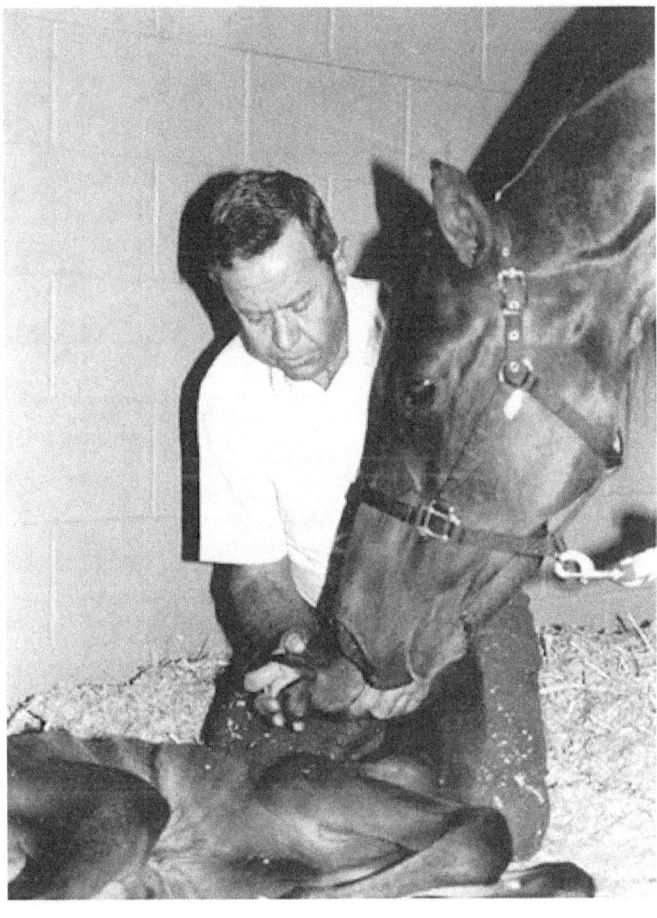
Testing desensitization of the upper lip. This completed the first session with the filly-it took about 50 minutes. I released her, she got to her feet, and was soon nursing.

self, even though I recommend that two people do it—one to hold the foal down while the other person does the procedures. This is especially important for novices at imprint training, or for anyone who lacks the strength or experience to prevent the foal from getting up.

Some foals are surprisingly strong, and an inexperienced handler can get kicked by the foal's flailing legs if care isn't taken. Working alone, I can keep the foal from getting up simply by keeping the head bent around to his upper side, as described.

When I was imprint training some newborn foals at a friend's ranch in order to get photographs for this book, the foal's owner made an important and, I think, significant observation. She expressed the opinion that flexing the foal's head and neck laterally, and maintaining that position for a minute or so at a time, probably facilitated the halter breaking which I usually do on the following day.

I hadn't appreciated it before, but I think that she is correct. Soon after birth, the foal learns to *yield* the head laterally, exactly what he will be asked to do the following day, when he is on his feet and when he is taught to lead. Also, if the foal happens to be a little stiff in flexing laterally, this helps to supple him. These might be two reasons foals learn to lead so quickly after being imprinted.

Similarly, as the foal is handled soon after he is born, after the legs are desensitized to touch, I repeatedly flex each leg joint, and then hold each leg in a comfortable flexed position for at least a minute. The foal

IMPRINT TRAINING

Now I move to the nostrils . . .

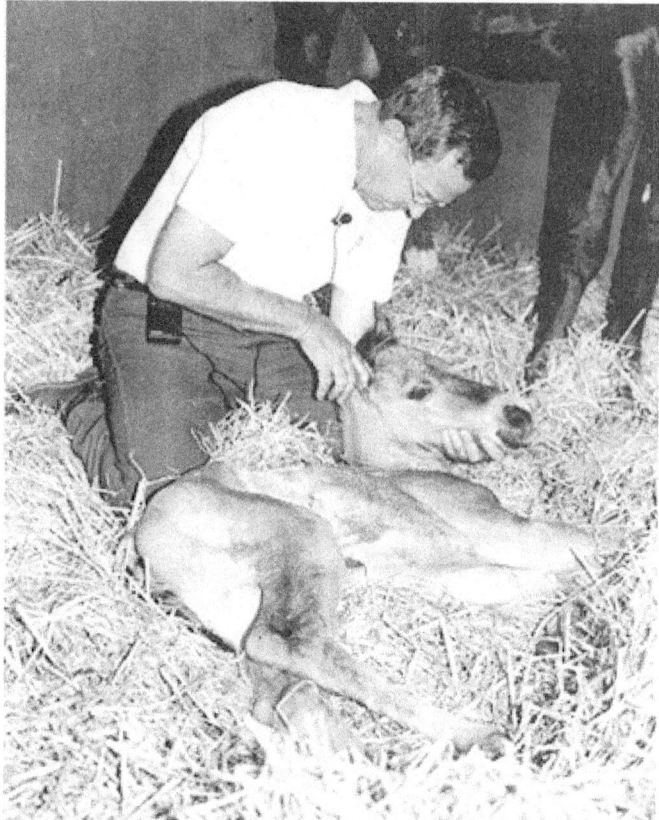

This next set of pictures shows some of the imprint training on a newborn foal at Malibu Valley Farms, which is a Thoroughbred farm. However, this is a Quarter Horse foal out of a Doc Bar mare, who was boarded at the farm for foaling and imprint training of her foal. Although the foal has not been on his feet yet, the mare has licked him and bonded with him. Here, I am desensitizing the ears.

. . . the upper lip . . .

IMMEDIATE POSTPARTUM PROCEDURE

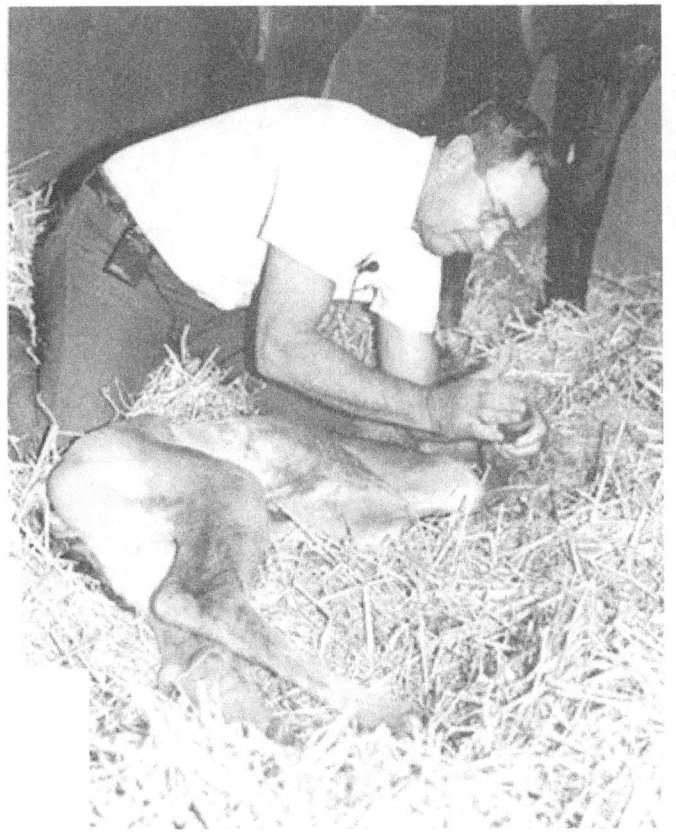

. . . a forefoot . . .

. . . the right rear leg . . .

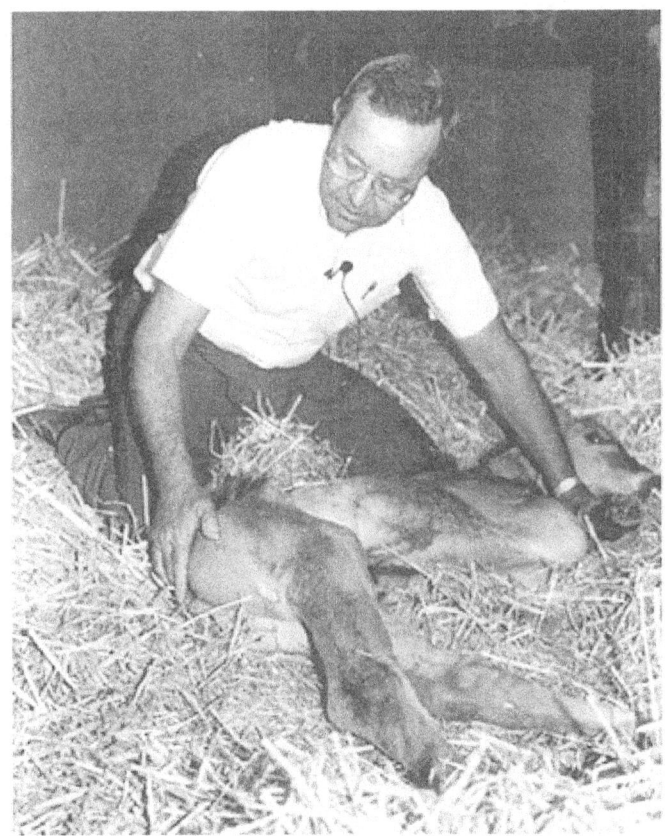

. . .and the perineum (under the tail).

usually struggles a bit, but soon relaxes and yields the leg. The next day, when I pick up a leg and hold it flexed, there is little resistance. The leg is surrendered compliantly.

A week later, I can usually put a loop of twine around any one of the four pasterns and, gently, with little effort, elevate that leg, and hold it for 20 seconds or so in a flexed position. These foals learn, before they are one hour of age and before they have even been on their feet or nursed, that it is futile to resist when a leg is picked up. They are, in effect, broke to having any one leg tied up or hobbled. They will not resist having their feet trimmed or shod, and are much less likely to panic if a leg is caught in a loose rope or in wire.

Learning, at birth, to yield the head and the feet also helps to create a submissive, willing, and dependent attitude in the foal toward humans. As discussed elsewhere in this book, *flight deprivation* is one of the easiest ways to establish dominance in horses. When deprived of flight, this species' primary survival behavior in the wild, the horse feels that he cannot protect his life, and seeks leadership that can somehow give him safety.

I have observed similar changes in attitude in other species when they are deprived of their primary survival behavior. Thus dogs, whose teeth are their principal weapon, are rendered helpless when muzzled, and cattle, whose horns are their principal weapon, are made helpless when their nose is elevated and their head immobilized.

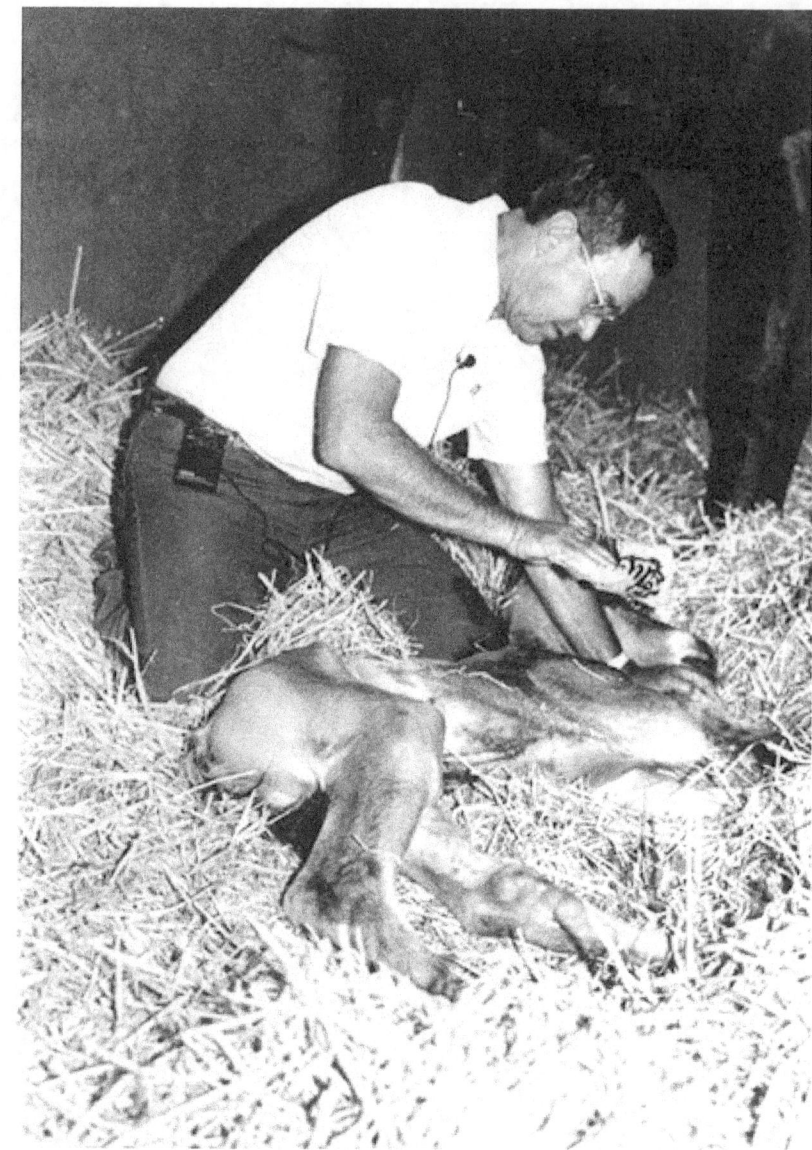

Desensitizing the foal to crackling plastic . . .

IMMEDIATE POSTPARTUM PROCEDURE

...and the electric clippers. I continue working with this foal in Chapter 8—after he's on his feet.

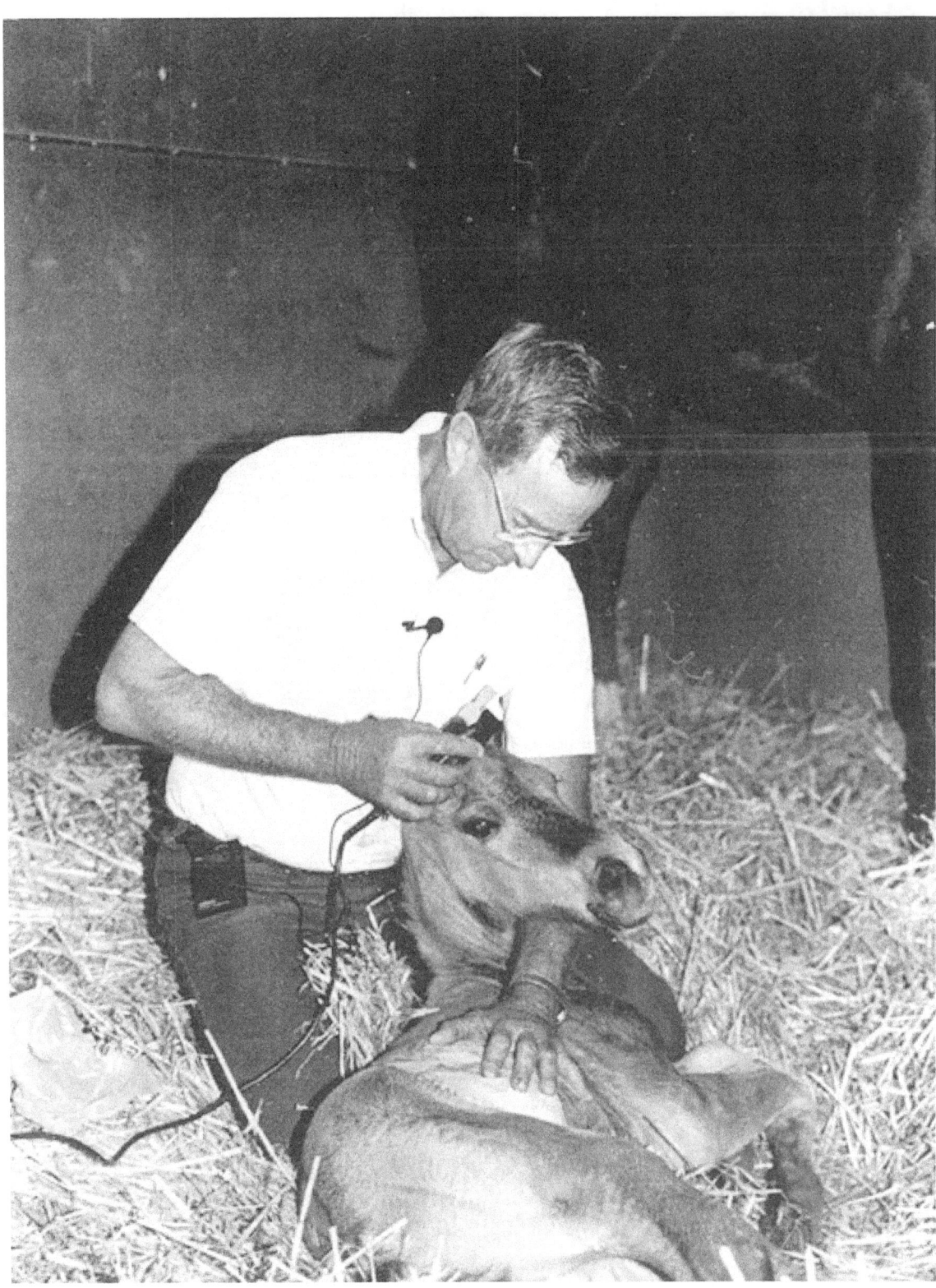

8 The Second Session

Photos by Debby Miller

THE SECOND training session can be done after the foal has nursed and is strong enough to stand up for a while. Newborn foals fatigue quickly and, unlike mature horses who can sleep standing and stay on their feet indefinitely, foals must lie down to rest. Therefore, this second stage of training should not exceed 15 minutes in length. If it is not completed, stop after 15 minutes and come back later when the foal is rested and complete it.

Sometimes a mare will surprise us and foal unexpectedly, so that when we discover the foal, he is already on his feet. In that case, all of the imprint training described in the previous chapter can be done on the standing foal, but instead of being done in one continuous hour-long session, it should be done in 15-minute increments. Alternatively, the foal can be laid down again, not allowed to arise, and desensitized as if he had not yet been on his feet.

Foals must be laid down gently. For the novice, I suggest waiting until the foal lies down by himself; then approaching him from behind—perpendicular to his spine. Keep him down by bending his nose toward his withers and flexing his upper foreleg.

In any case, there are several desensitization procedures which are easier to do after the foal is standing. The first of these is to accustom the foal to pressure on his back. Ideally, a team of three people

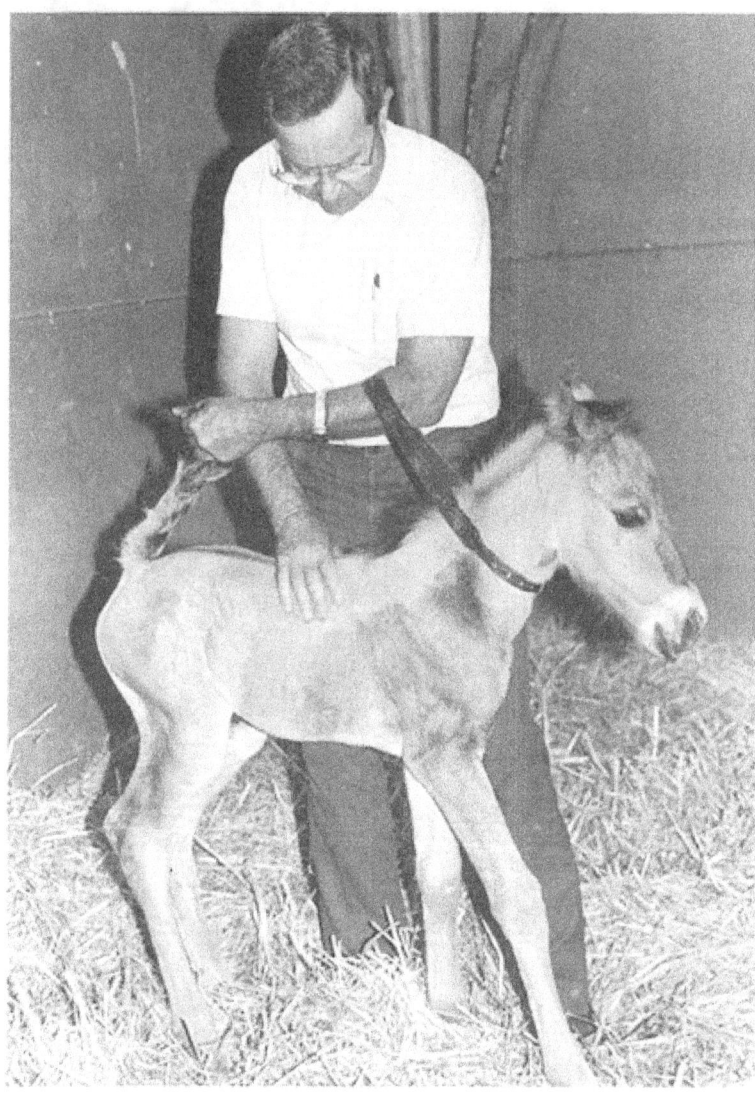

Starting the second session with the Quarter Horse colt that I began working with in Chapter 7. He is now on his feet, and has nursed. I use my belt to support the forequarters and a tail hold to support the hindquarters while I desensitize the saddle area to pressure. It is better to do this with two people—one to hold the foal and the other to desensitize him—but I am used to working alone.

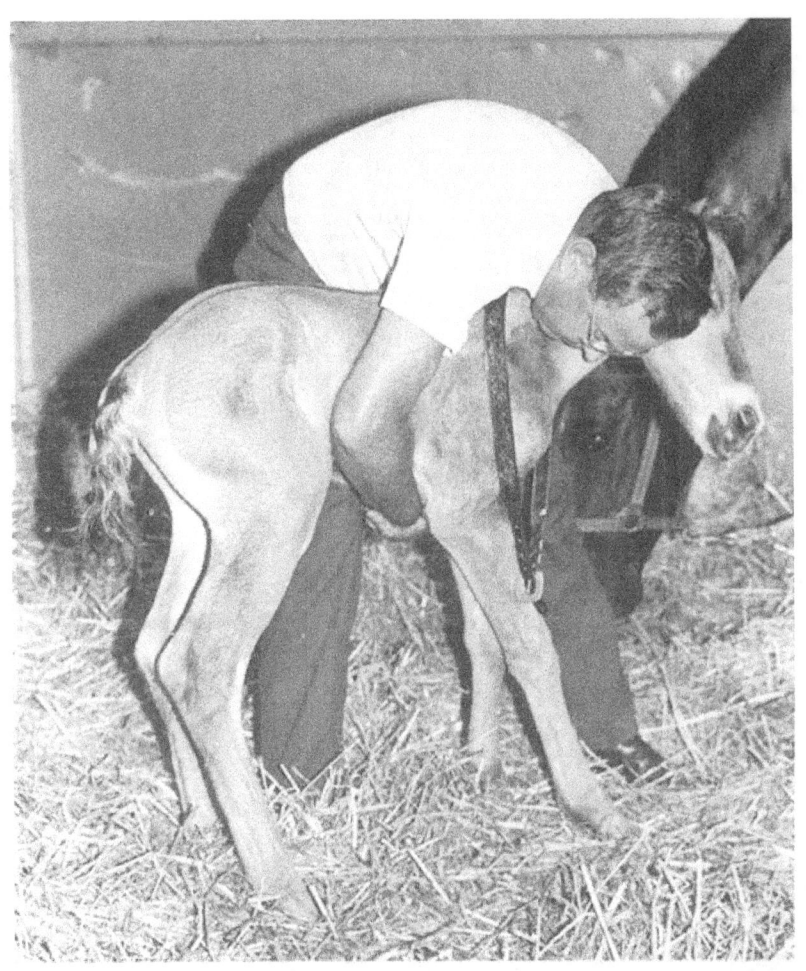

There are several desensitization procedures which are easier to do after the foal is standing.

should be used. One holds the mare, a second holds the foal, and the third does the actual desensitization procedure.

Using the flat of the hands, press the saddle area of the standing foal just hard enough so that he feels the weight.

Repeat this rhythmically until habituation occurs. Then encircle the girth area with your arms, and, with hands clasped under the foal's chest, rhythmically and repetitiously squeeze the foal at least 50 times. This will desensitize the foal to girth or cinch pressure, which is the primary reason horses buck when saddled for the first time. This is a procedure which simply cannot be done competently while the foal is lying down.

I like to repeat this squeezing process with my arms around the flank area of the foal just in front of the hind legs.

Now pick up each foot and tap the sole. This should not be a problem if the initial desensitization was done properly. The foal should now yield the leg in a relaxed manner and ignore the foot being tapped or rubbed.

If, while we are doing this second session, a well-fitting halter is put on the foal, he will not be disturbed by the presence of a halter at future training sessions. Having seen many foals crippled or killed by getting their halters hung up, I never leave a halter on an unattended foal, however. Imprint-trained foals are easy to catch, so there is no reason to leave a halter on them.

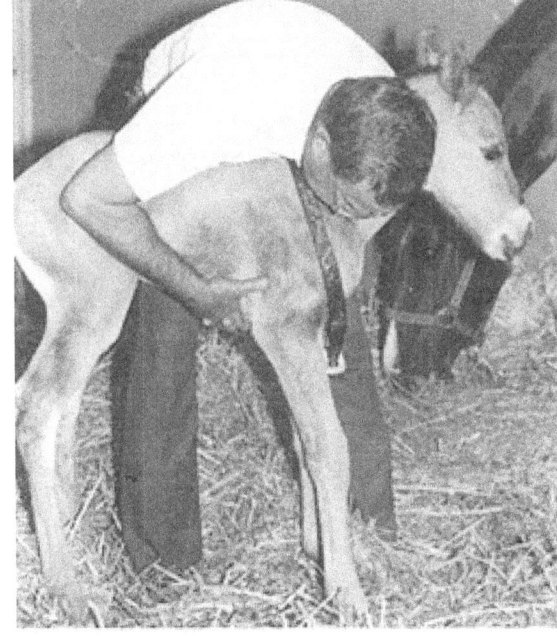

These two photos show how I desensitize the girth area to pressure. I squeeze this area, rhythmically and repetiously, about 50 times.

IMPRINT TRAINING

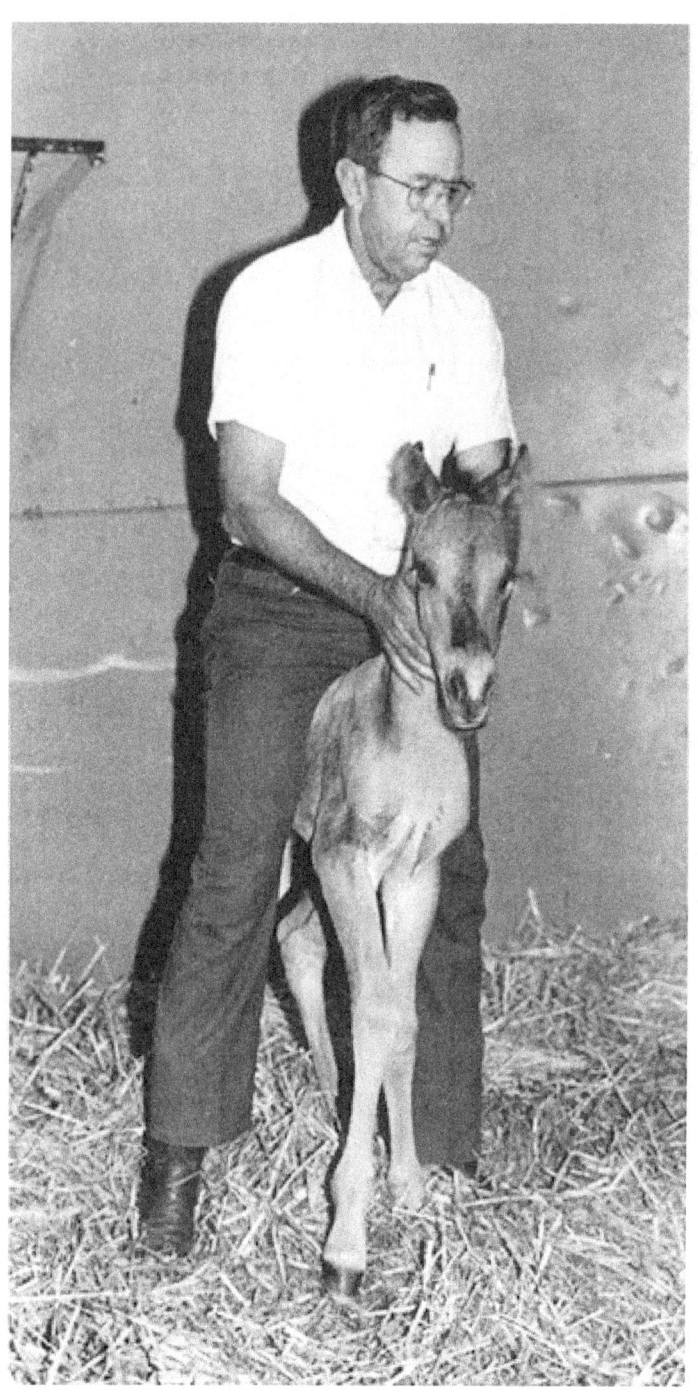

Straddling the foal. He will recall this the first time a rider mounts him. Straddling the foal should only be done if you are tall enough not to put any weight on his back.

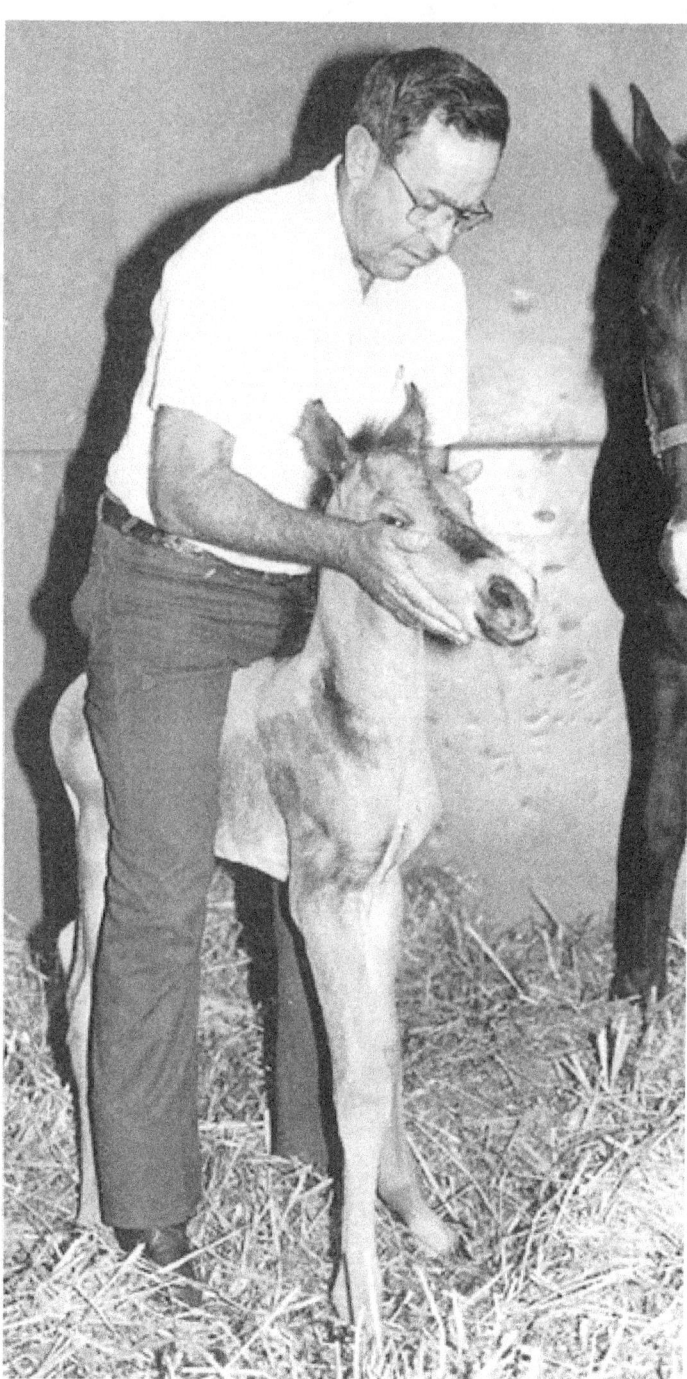

Rubbing the face to continue desensitizing it while I straddle the foal.

THE SECOND SESSION

Sensitizing the foal to back up in response to chest pressure.

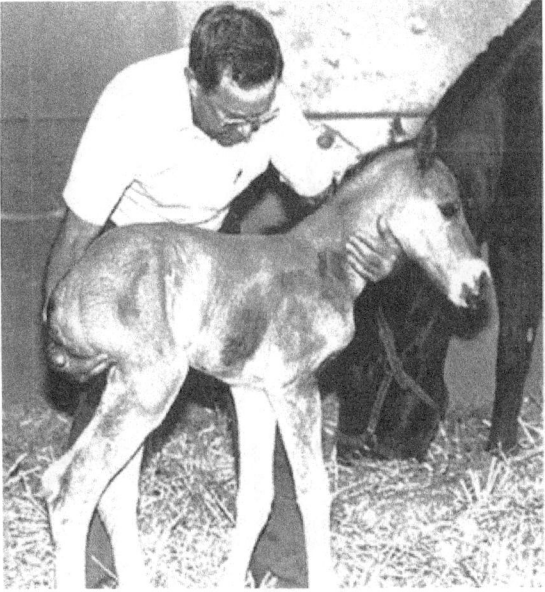

As soon as he takes a step back, I move my hand away from his chest. It only took three times for him to learn to move back from chest pressure.

Foals should be desensitized, as well as sensitized, to pressure.

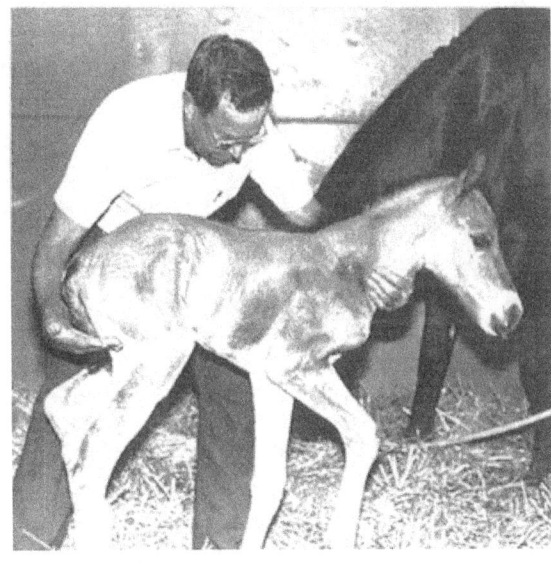

Three times did it.

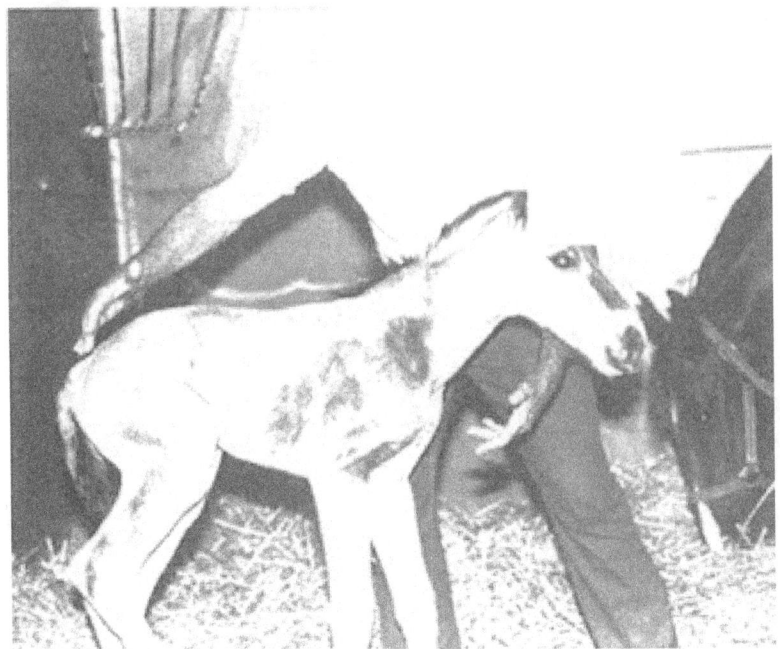

Sensitizing the foal to move forward in response to butt pressure

IMPRINT TRAINING

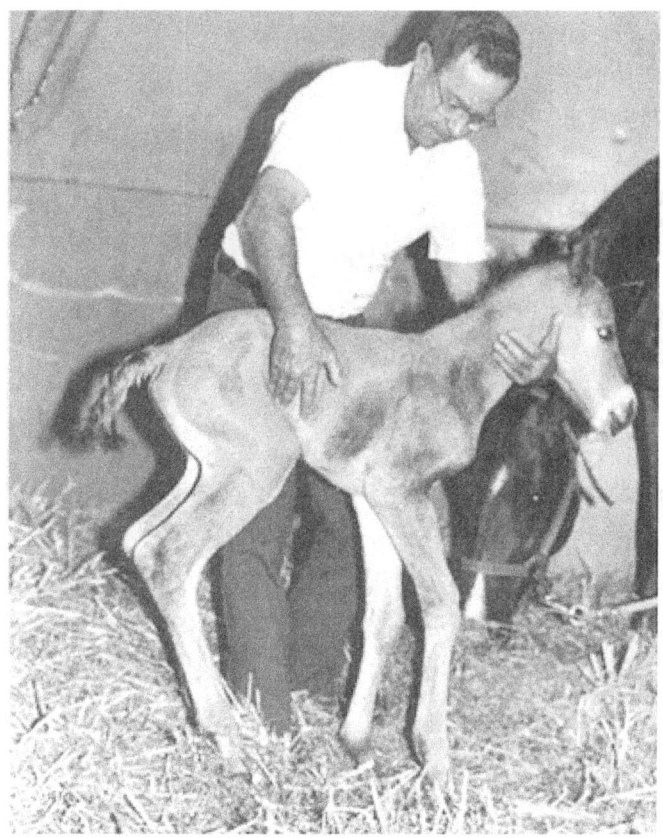

Sensitizing the foal to move his hindquarters toward me—in response to flank pressure.

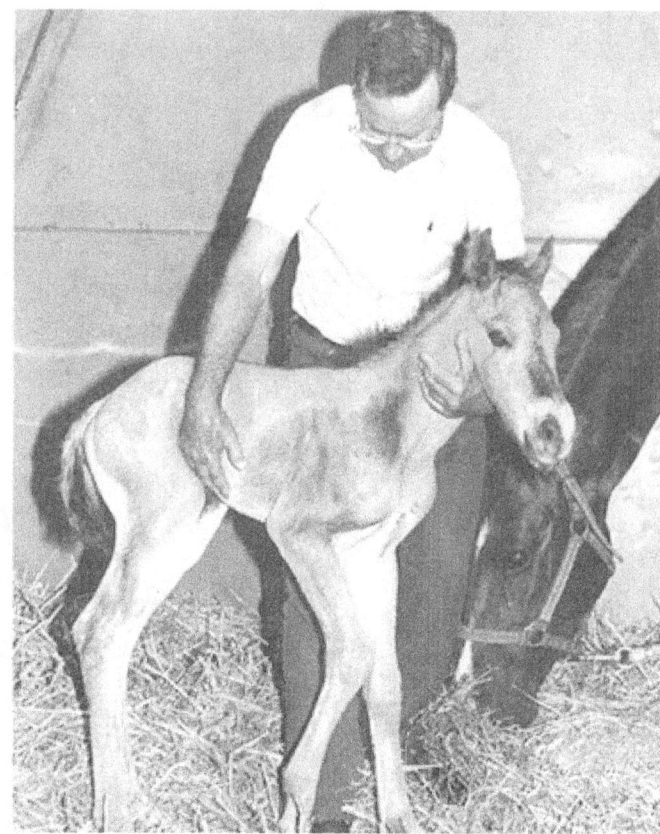

Three times and he learned to do it. I repeat it a few times to reinforce the response.

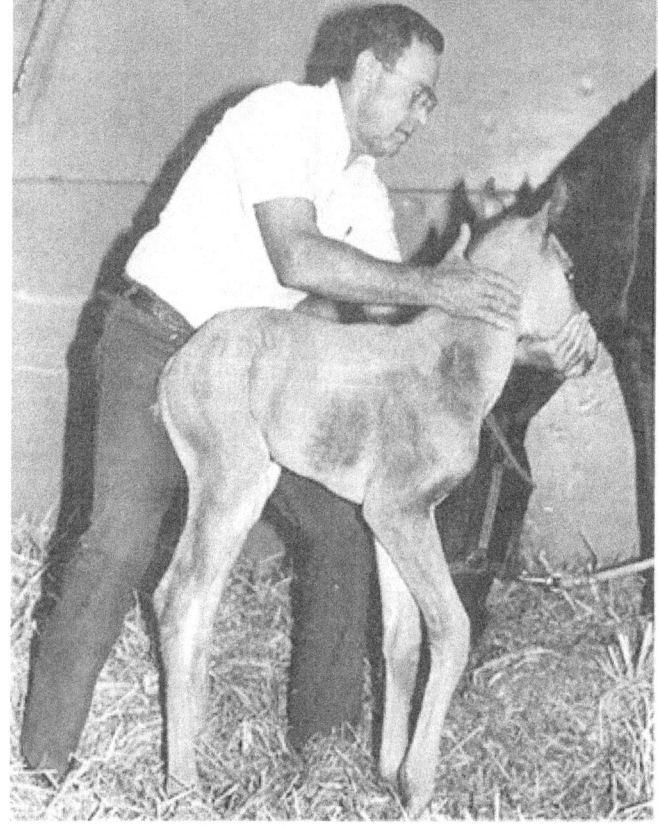

Sensitizing the head and neck; the forehand moves toward me. I only ask for one step, and he learns the desired response after three repetitions. My hip holds the hindquarters stationary.

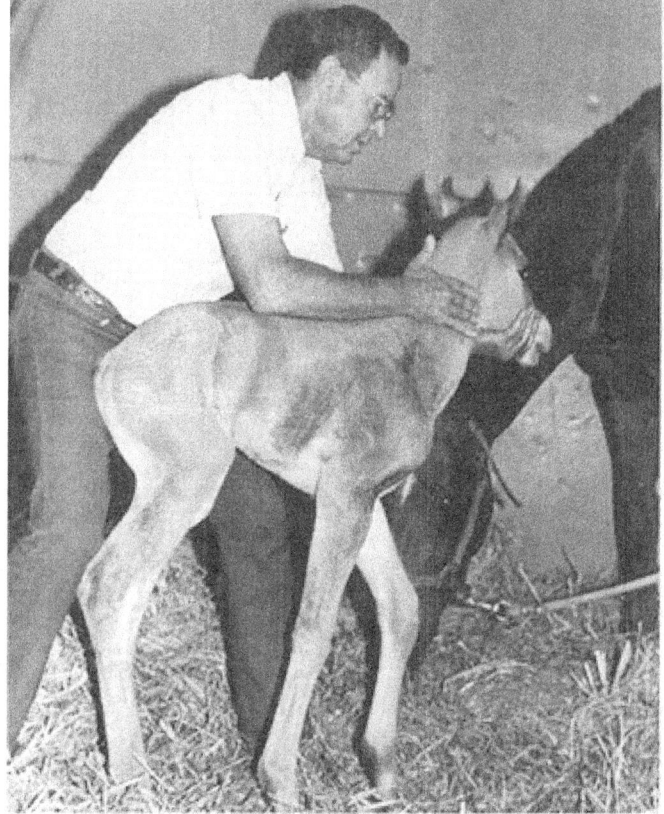

Turning the forehand to the left while the hindquarters remain stationary.

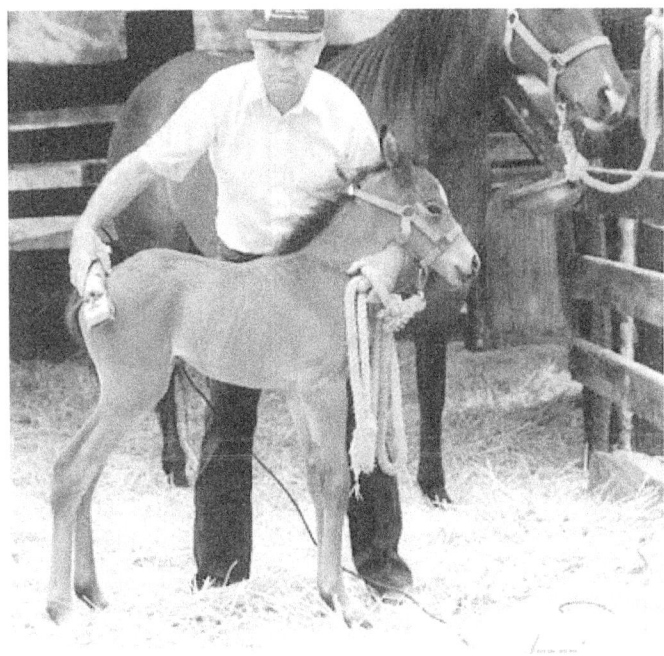

Here are four pictures showing how I desensitize a foal to electric clippers. Ideally, this should be done within the first hour of foaling, but there was no electricity where this Quarter Horse/Thoroughbred filly was born, so this procedure was not done until the next day, when the filly was 26 hours old.

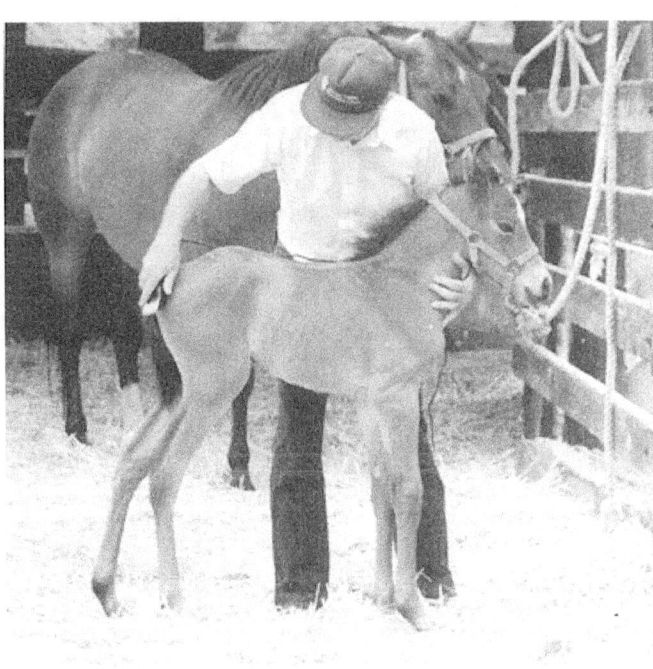

Since the filly was imprinted at birth, however, using the clippers went smoothly.

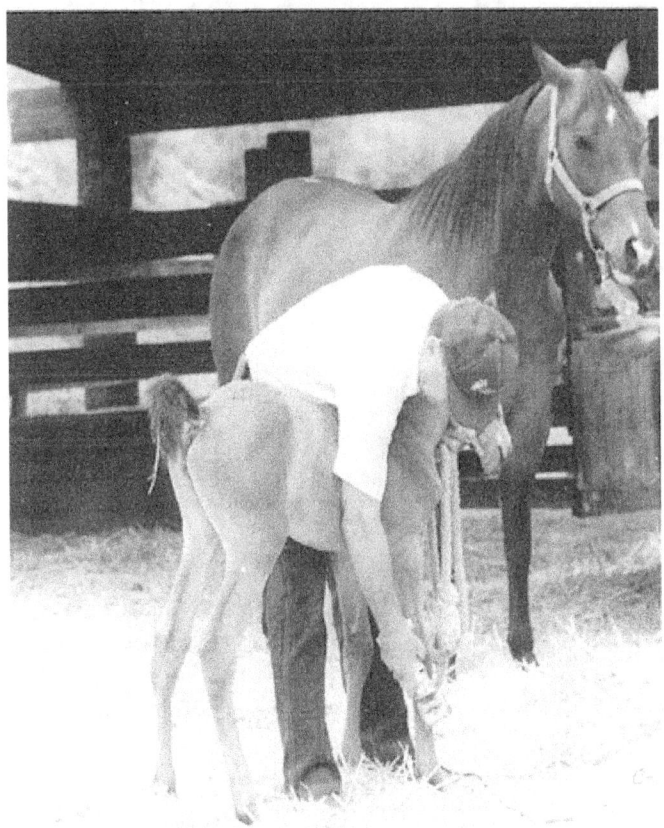

I even move the clippers up and down each leg.

I am not actually clipping her; I am simply accustoming her to the sound and vibration of the clippers.

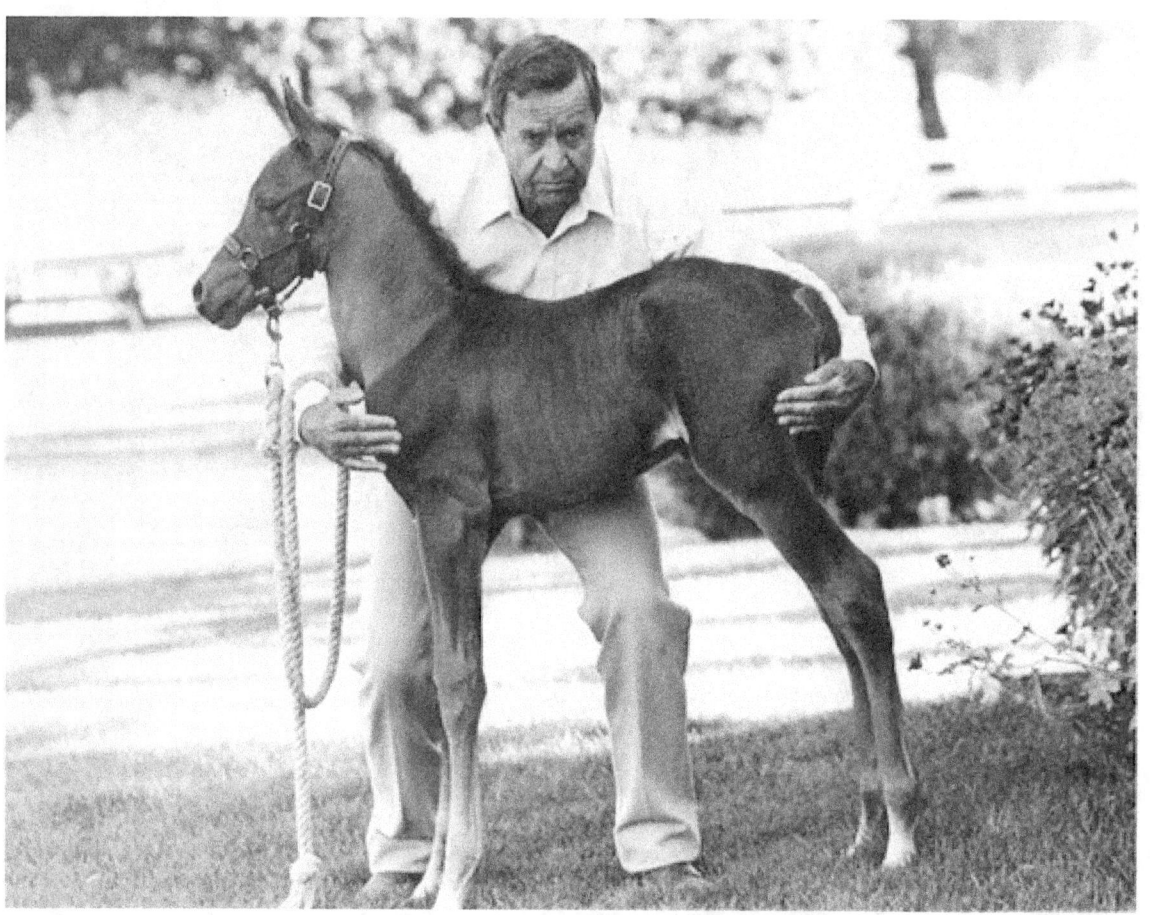

Stopping while the foal is resisting will teach him to resist in the future.

This Arabian colt, owned by Ventura Farms of Thousand Oaks, Calif., had 1 hour of imprint training at birth. He was 36 hours old when I started the second session, and put a well fitting halter on him. Because he is not halter-broke, and because I don't want to grab and panic him, I make a "little corral" with my hands to confine him.

At this point, it is a good idea to put on a surgical glove, lubricate the forefinger with some lubricating jelly or Vaseline, and gently insert the finger into the anus. Wiggle the finger and the foal will soon be desensitized to this stimulus. It will be much easier to take his temperature for the rest of his life. If rectal palpation is ever necessary, the procedure will be greatly facilitated. Also, this foal will not be likely to get upset if a rope gets under his tail, or when he experiences a crupper for the first time.

Remember, when doing this—as in all habituation procedures—to persist in the stimulus beyond the point of acceptance. Stopping while the foal is resisting will teach him to resist in the future.

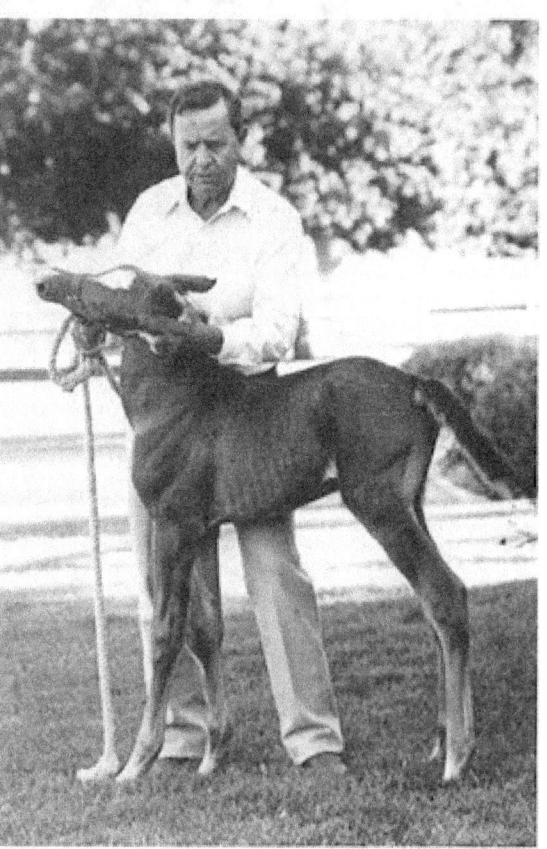

Testing desensitization of the left ear . . .

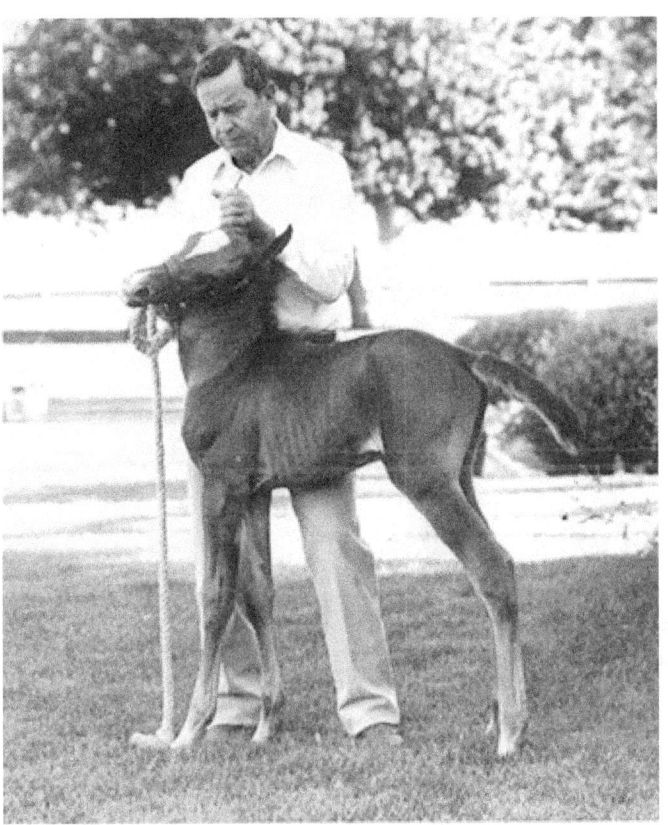

... and of the right ear ...

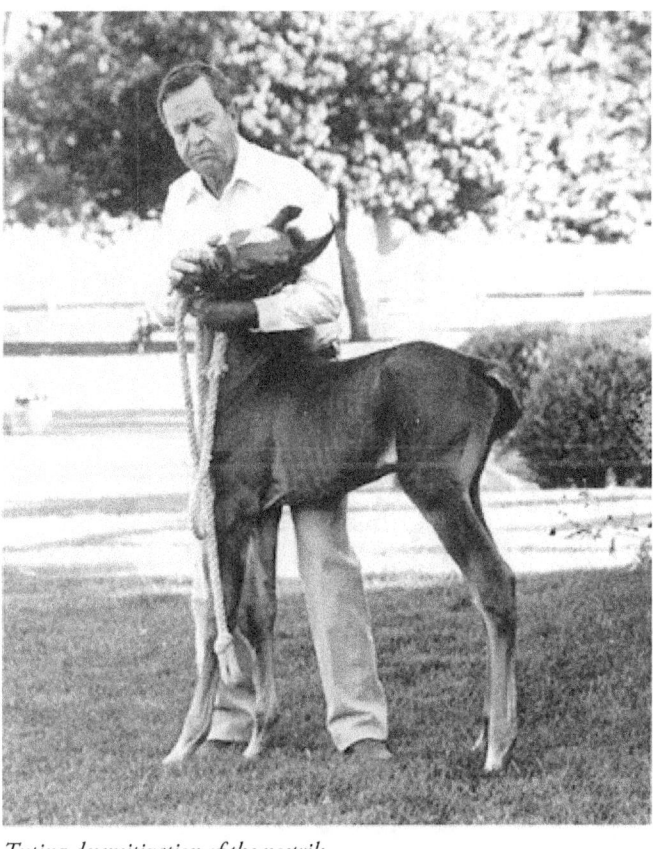

Testing desensitization of the nostrils.

... of the mouth and tongue ...

... of the saddle area ...

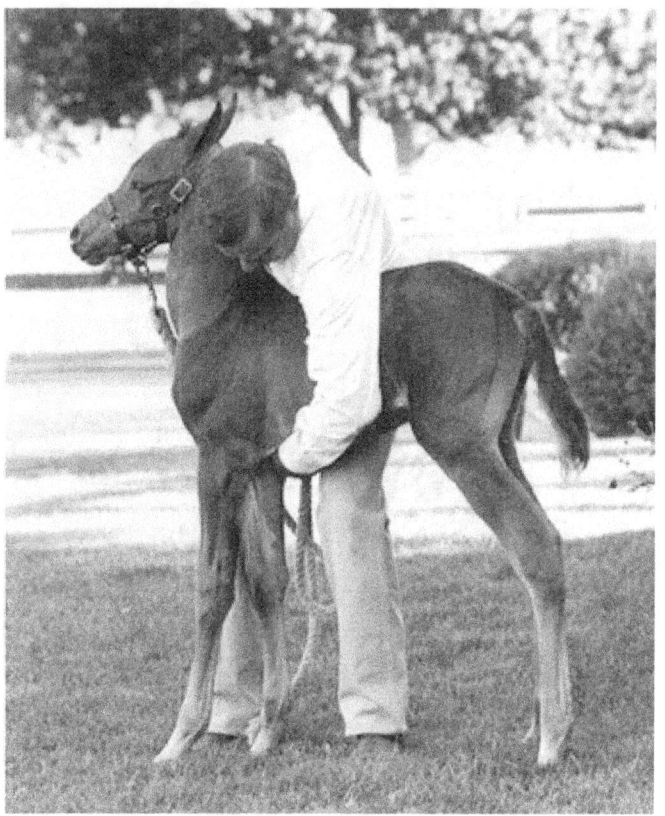

... of the cinch area ...

... of the perineum and tail ...

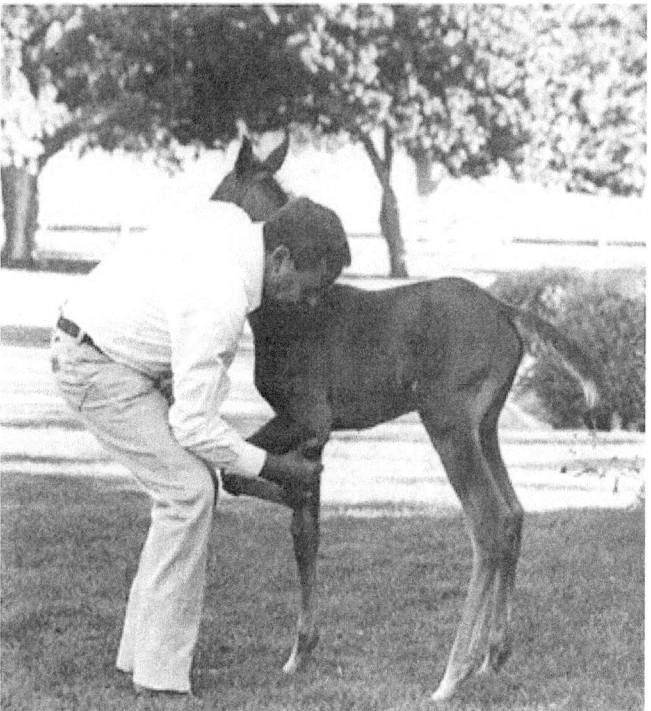

... and of the left foreleg.

Slapping the bottom of the foot to simulate horseshoeing.

THE SECOND SESSION

The colt is totally relaxed and lets me hold his foot with my finger tips.

Checking to see if he will allow me to hold his left hind foot.

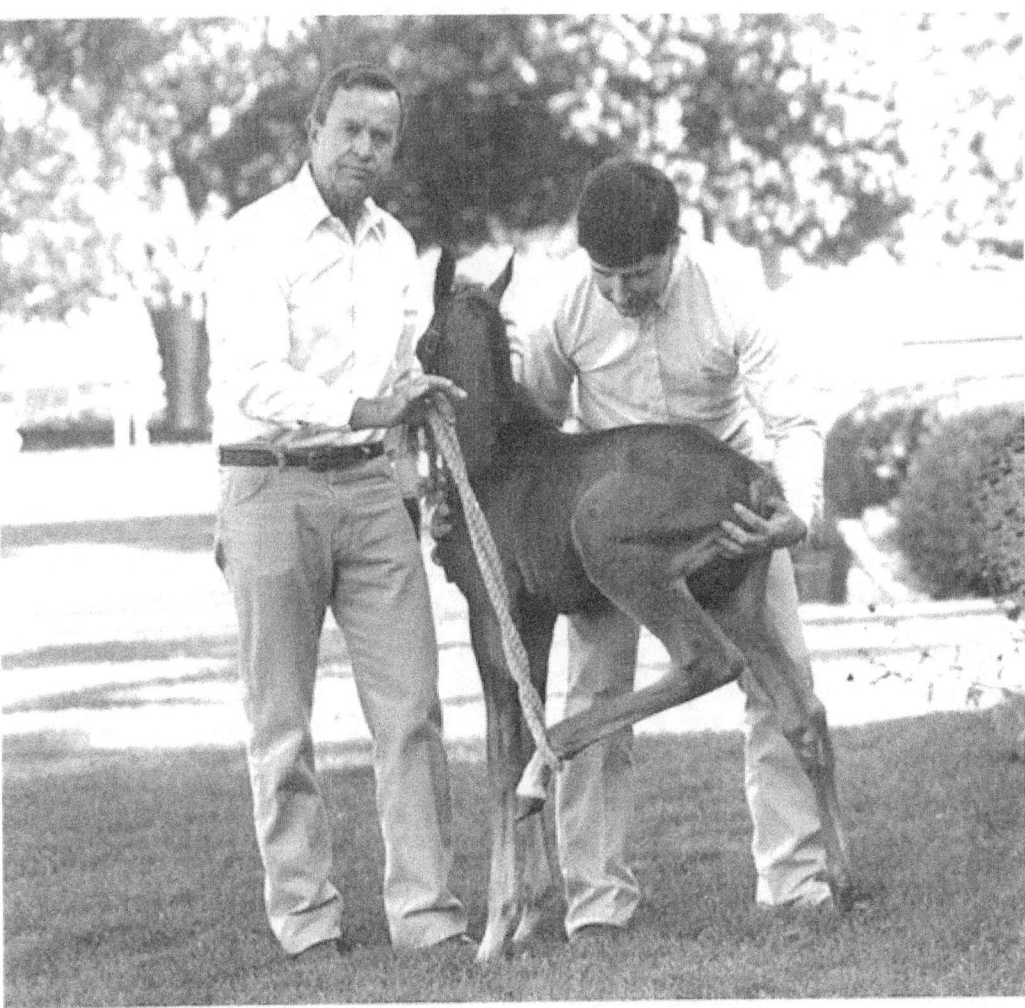

I'm doing the same with a rope around the pastern. It was effortless!

9 THE THIRD SESSION

Photos by Debby Miller

THUS FAR, we have spent about an hour and a half training the foal, and our efforts have been mostly directed at desensitizing him to frightening stimuli. While doing this, we also have obtained the benefits of having done it during the imprinting period. The foal will therefore be bonded with us and will recognize us as being dominant to him. The relationship of trust and respect without fear will have been started.

The third session will primarily be used to *sensitize* the foal, not *desensitize* him. However, we will also briefly use it to reinforce some of our previous desensitizing procedures.

The time to do the third session is as soon as the foal is strong on his feet and well coordinated. In some cases this may be as early as 12 hours of age. In other cases it may be as late as 36 hours of age, and in the case of some very weak foals, even later.

The third session will primarily be used to sensitize the foal, not desensitize him.

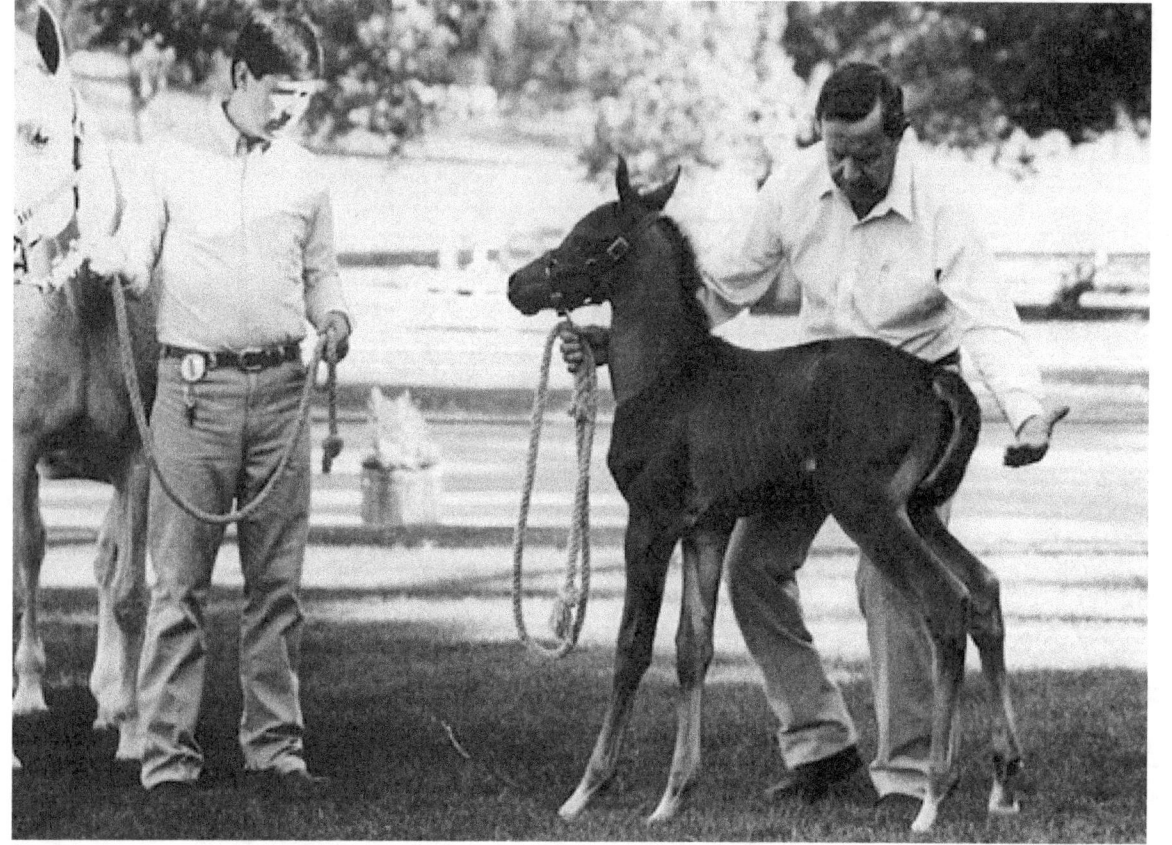

In the previous chapter, I worked on desensitizing this foal. Now, I am beginning sensitization procedures, and here I'm asking the foal to step forward in response to pressure on his butt. As soon as he took a step forward, I released the pressure.

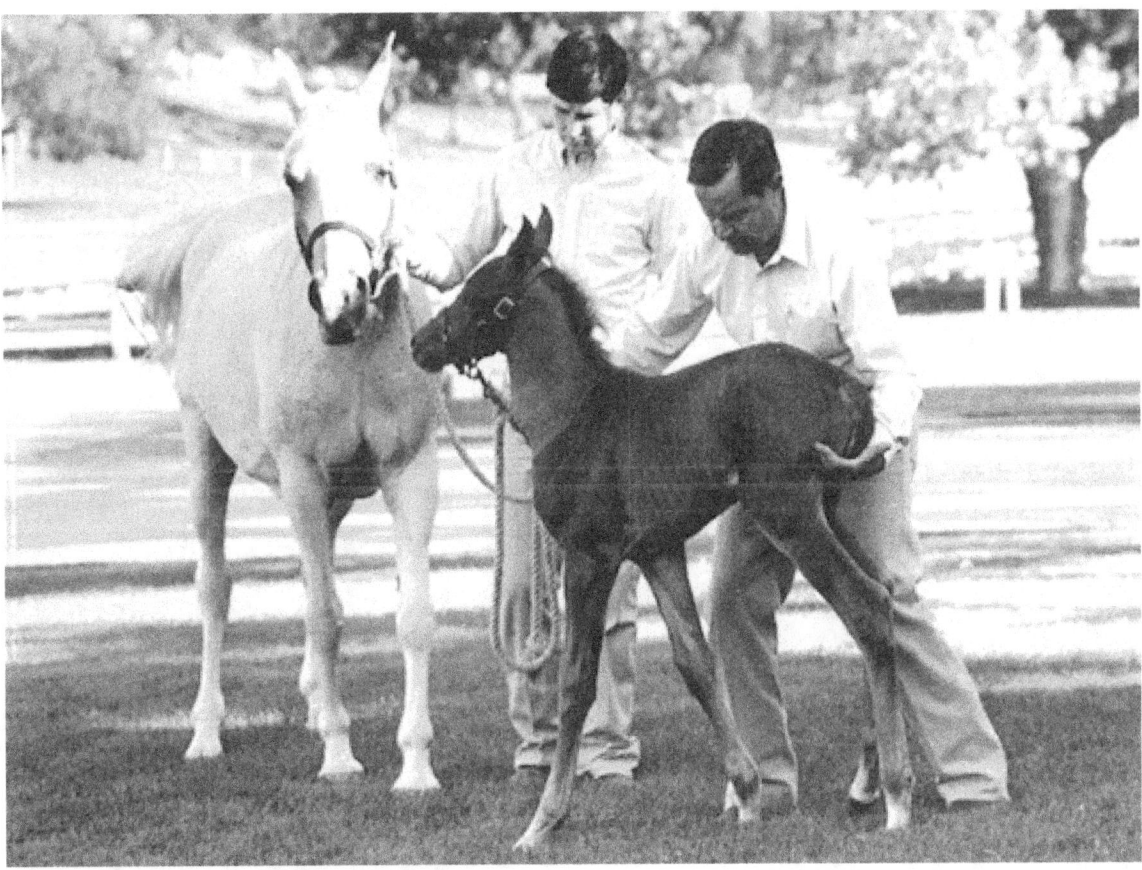

ere, I've asked him to take another step forward. Dr. Tim Evans, resident veterinarian at Ventura Farms, is holding the mare.

Now I am teaching him to back up in response to chest pressure. He took one step.

Typically, most foals are strong enough at 24 hours of age. Generally speaking, mule foals and horse foals out of strong, athletic mares with good muscle tone will be ready for the third training session in less than 24 hours, whereas foals out of flabby, obese mares, who have had little exercise during the gestation period, will be too weak and wobbly to work with until after the first day. Foals that are down in the pasterns with the point of the fetlock close to the ground or, conversely, foals that are "contracted" with excessively steep pasterns, or foals whose legs cross or are knock-kneed or bowlegged are probably too uncoordinated to start this phase of training.

I personally like to do these procedures working alone, although it is nice to have an assistant holding the mare. However, a team of three can again be used with one mare handler and two people working with the foal. For the novice, an extra person holding the foal, at least for some of the pro-

He learned to back up after just three tries.

Repeating the backing lesson. One word of caution: Do not position your head directly above the foal's head. If he throws his head, he could crack you on the chin.

cedures, might prevent things from going wrong.

To begin with, it is a good idea to briefly repeat all the desensitizing steps already established: rubbing the foal's head, body, legs, etc., and picking up each foot, tapping the soles, and inserting fingers into the various body openings. This will let us know if any area is not adequately desensitized, and will reinforce the habituation to those that are. All of this can be quickly done in a few minutes.

Now we are ready to sensitize the foal, which will include halter breaking and other maneuvers. Proper halter breaking, regardless of the age that it is done, is the foundation for all future training. It is the most basic and most important part of all training. Most horses are halter-broke, but not all of them are properly halter-broke.

The properly halter-broke horse, when tied, will not pull back. He believes that once tied, it is impossible to escape. Therefore, he will not try. Because he believes that escape (flight) is impossible, simply putting a halter on him will immediately create an attitude of submissiveness.

When led, the properly halter-broke horse will obediently and submissively fol-

Now I'm teaching him to move his rear end sideways (toward me) in response to flank pressure. I restrain the forehand from moving. He is also bending around my right leg—as he will do later in response to the rider's leg aids.

He did it. I only ask for one step at a time.

Learning To Back Up and Move Forward

Standing alongside the foal, press your fingertips against the base of the neck just above the chest. Eventually the foal will step backwards, or at least lean back.

Reward instantly by abruptly stopping the pressure. If you do this correctly, in just a minute or so, the foal will back up in response to this pressure.

Now repeat the above step, using the hand to create pressure on the hindquarters, below the tail. The foal will soon move forward when he feels light pressure from behind. Eventually we will be making a transition from the hand to a butt rope to encourage the foal forward when led by the halter.

Gradually expand the circle. The foal is now leading. If he hesitates, a butt rope will stimulate forward movement, providing the foal has been previously conditioned with hand pressure.

Now I do the other side.

low. He believes that escape is impossible and therefore he can be led with a piece of string. Of course, a severe fright can cause any horse to try to run off.

This is the response I want.

I repeat the cue several times to "fix" the response.

Learning To Move the Hindquarters Laterally

The next objective will be to sensitize the area where the rider's heel will contact the side of the belly. Stand alongside the foal and place one arm under the foal's neck in order to restrain him (see photos). Then reach over the foal's back with your other arm and press the fingertips into the flank. Maintain pressure until the foal moves toward you. The instant he does, move your body away from the foal so that you do not obstruct his movement.

Simultaneously and abruptly, snap your hand away from the flank. Pause. Repeat the procedure. In a few minutes the foal will have been conditioned to move toward you anytime he feels pressure on the opposite side. Do not ask for more than one step at this point. Repeat the training process from the other side. Your foal should now be conditioned to move his hindquarters away from pressure while his forefeet remain in place.

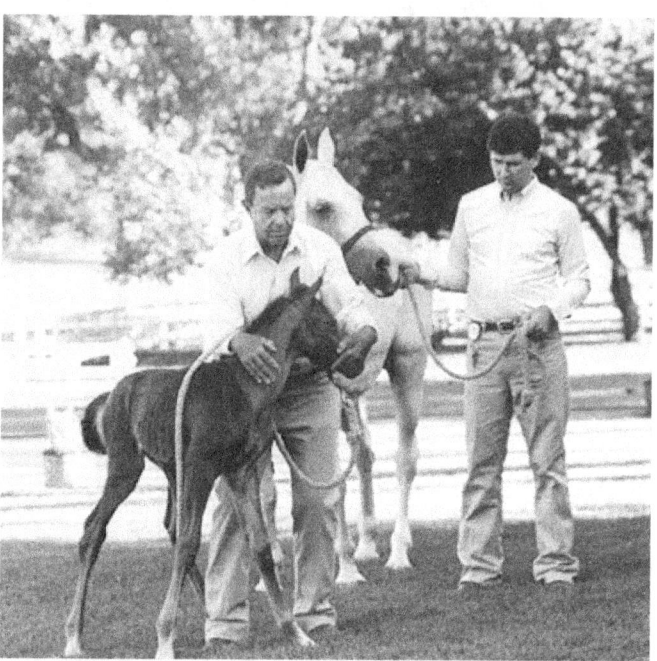
Next, using gentle hand pressure on the right side of the face and neck, I bend him to the left. This can be used as a preparatory step in teaching him to lead.

By way of explanation, I initially use flank pressure—behind where a rider's heel will eventually apply pressure—because it gives me more leverage. Later, I

I repeat it several more times.

Anytime the foal shows fatigue, stop the lesson.

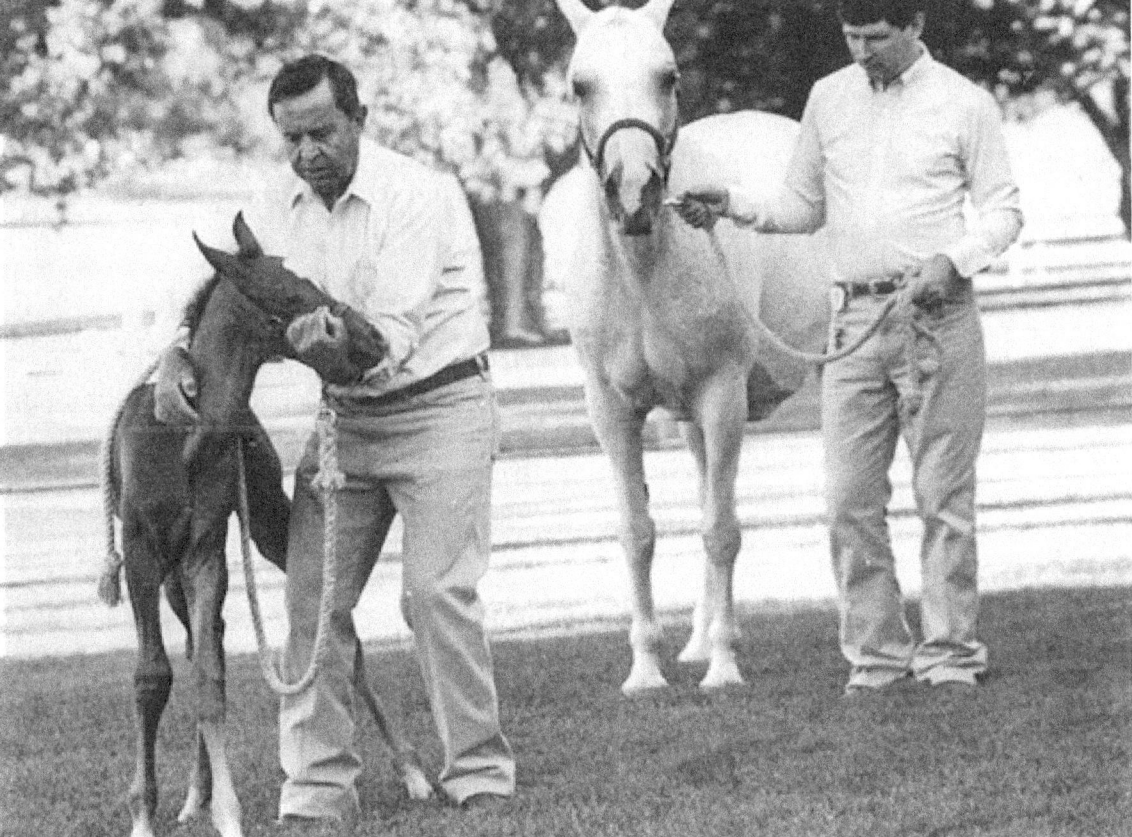

Bending his head and neck to the right.

gradually move the pressure forward a little at a time, making a transition to the area of a rider's heel contact.

During subsequent reinforcement lessons, you can begin to ask for two steps laterally, progressively shortening the length of the pause between the signals.

Still later you can ask for three, and then four separate movements in the same direction. In time, your foal will rotate on his forehand, moving the hindquarters laterally in a complete circle if the pressure is maintained. The importance of this step is second only to halter breaking (teaching to lead and tie). Too often trainers fail to gain such complete control of the hind quarters, and here we have obtained it in the baby foal.

As the foal grows, it will become more difficult to reach over his back to apply pressure, especially if the foal is tall and the trainer short. No matter. We have, by using an arm over the back, simulated the rider's leg when astride, our fingertips representing the heel and/or spur. Once the foal is conditioned in this way to automatically respond to the signal by moving the hindquarters laterally, we no longer need to reach over the back. Simply touching the belly on the same side as we are standing will elicit the same response. The foal will move away from the pressure.

In addition to obtaining maneuver ability, gaining hindquarter control at an early stage has another important value. In order to sprint away from perceived danger (flight), the horse must have his hind quarters under him and his spinal column aligned, just as a human sprinter does in the starting block. Lateral displacement of the hind end interferes with flight. Thus, once again, we have found a technique which prevents flight and thereby creates submissiveness in the horse. When the hindquarters are moving laterally, it is difficult for a horse to sprint, to rear, or to buck.

So, whether the handler is on the ground or in the saddle, control of the hind end gives him control of the entire horse. Forward motion comes from the hind end. The hind legs and their powerful musculature propel the horse forward. Upward motion, as in bucking, jumping, or rearing, is simply forward motion redirected upward. The forequarters of the horse control the direction of propulsion.

It is much like a motorcycle. The front wheel can be turned to direct which way the bike will travel. If a wheelie is done, the bike will "rear" The back wheel is fixed, but it is the source of power. If the back wheel skids to the side as the wheel rotates, instead of going forward, the bike will rotate onto its front end.

That is comparable to what happens when, with leg pressure, we divert the hind end of a moving horse to the side. Forward motion is lost. Thus, the reins, controlled by our hands, serve as the "handlebars" to our analogous motorcycle, while the hind limbs, controlled by our legs, serve as the powered rear wheel.

Both the forequarters and the hindquarters of the horse can be conditioned to respond to our signals in a foal that is a day or two of age, and these responses will be retained and enhanced if we periodically repeat and reinforce the procedures.

Learning To Lead

Horses can be properly halter broke at any age, but in this book we will limit our discussion to the newborn foal. Be sure that the foal's halter fits properly, neither too large nor too small. It should be strongly made. The foal should stand roughly parallel to, or in front of, the mare, his head facing in the same direction as the mare's. Place yourself at the foal's side, between mare and foal. With one hand holding the cheek of the foal's halter, slowly pull the foal toward the mare. As the head is drawn to the side, the foal will eventually have to move the foreleg on the same side towards the mare in order to maintain his balance. The instant that leg moves, no matter how slightly, abruptly release pressure on the halter to let the head return to its normal position, in line with the body, but do not let go of the halter. Wait a minute and then repeat the procedure.

If the reward (relief from pressure) is swift enough, and this depends upon your perceptiveness and the speed of your responses, after just a few times, the foal will step to the side in the direction that his head is being pulled. The foal is already learning to follow the direction in which the halter is leading it. Ask only for one step. Even a half of a step should be immediately rewarded.

Move to the other side of the foal and repeat the same procedure. Soon the foal will step to the side when the halter is pulled in that direction. Do not rush the procedure. With experience, the foal can be trained to follow the halter in just a few minutes, but why hurry?

Anytime the foal shows fatigue in this or any subsequent lesson, stop. The training can be resumed after the foal has nursed and rested.

Once the foal responds quickly to the halter with one step, pause, and ask for a second step in the same direction. Be satisfied with the smallest response, and reward every effort with a rest pause and a bit of petting. Soon the foal will be rotating on his haunches, following the halter by moving his forehand laterally in a small circle. Later, expand the circle and the foal will now be leading, but not yet in a straight line.

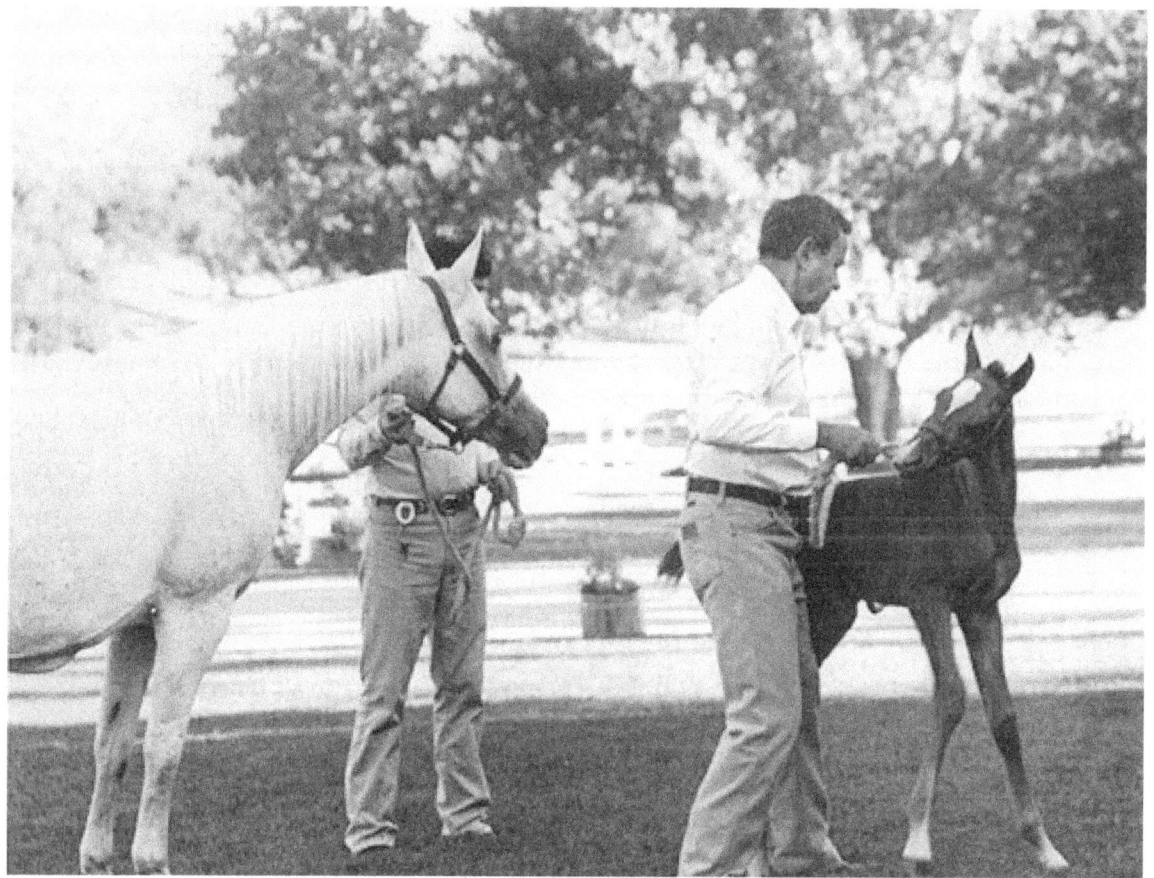

Now that the colt has learned to yield to head pressure, I gently pull his head to the side until he moves his inside (right) front leg toward me.

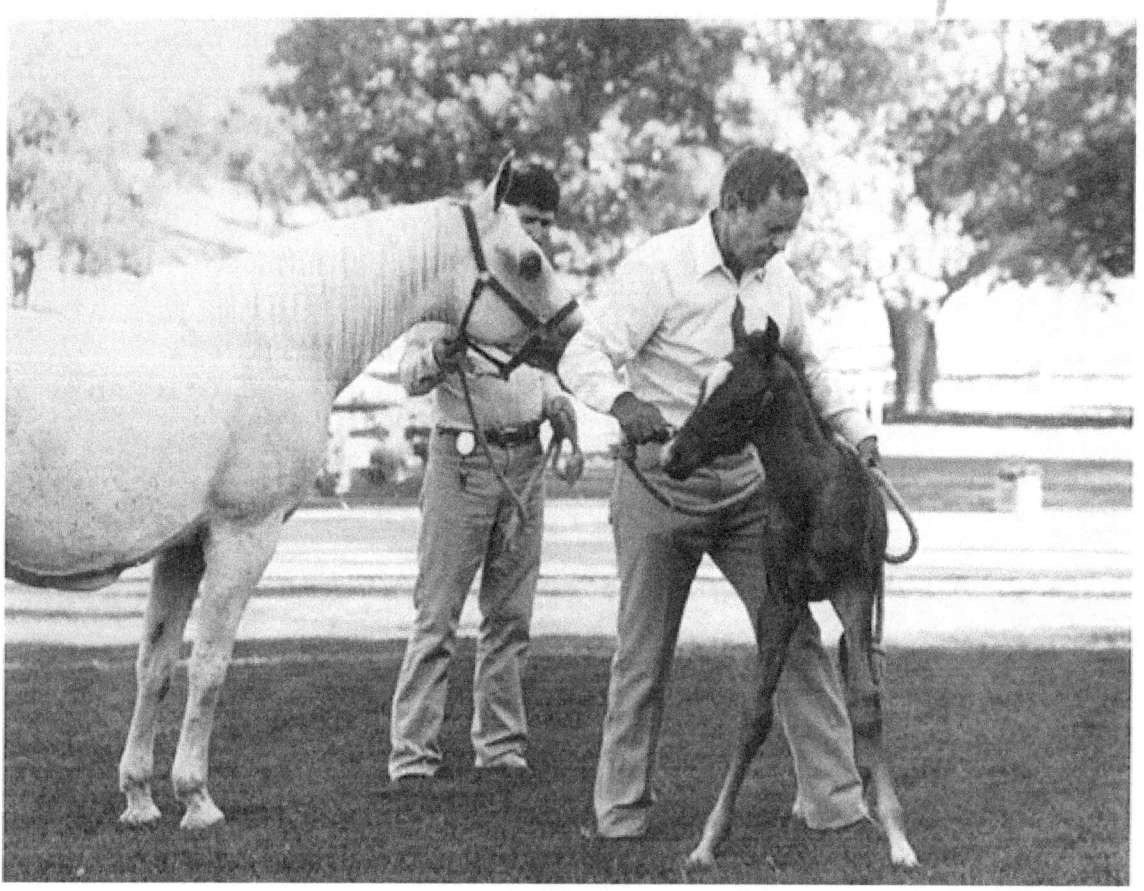

My left leg blocks the hindquarters, much as the rider's leg will later on.

In 5 minutes, the foal can be led in a small circle, almost rotating on his hindquarters. He never blows up or gets excited because of the foundation work I have done.

I repeat the same procedure to his left.

IMPRINT TRAINING

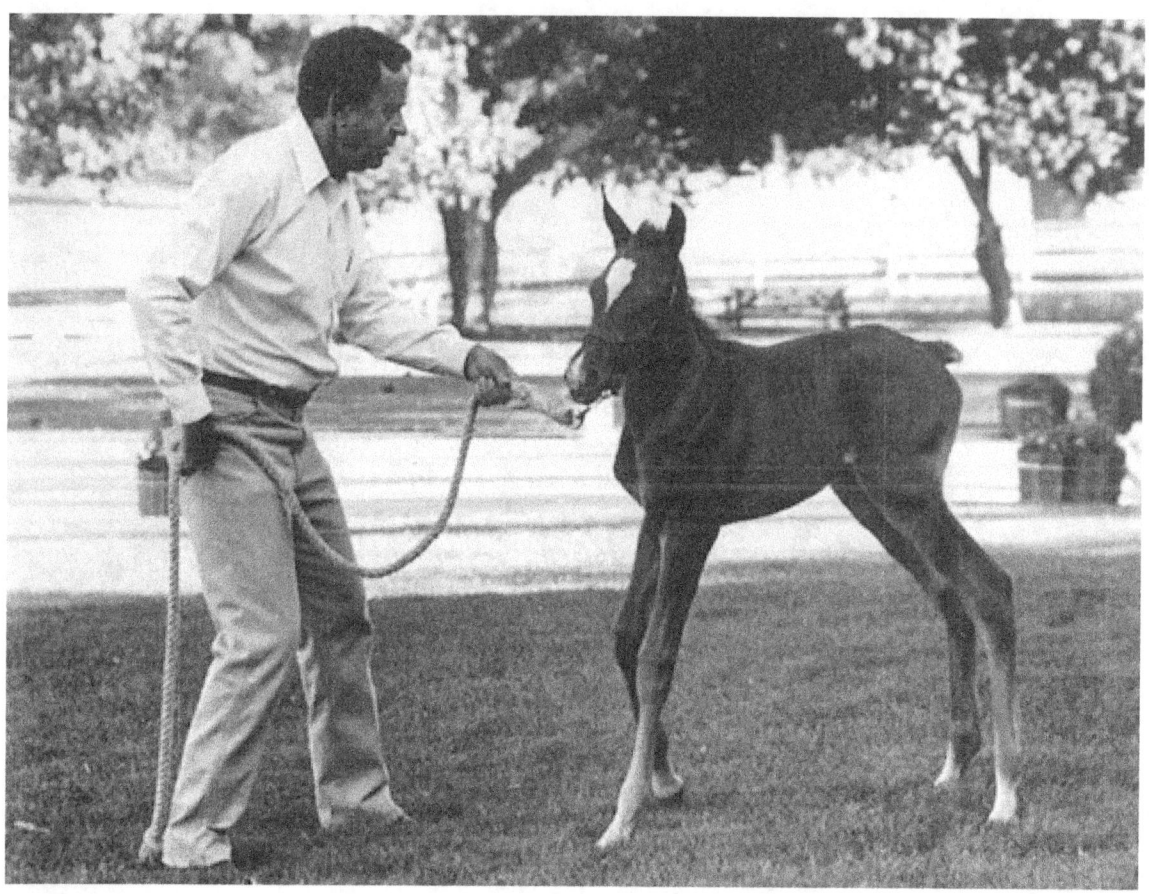

I expand the size of the circle, moving farther away from the mare.

Now I'm teaching him the basics for standing tied. I stand very still, and gradually and gently pull him toward me. Often a foal will hop or jump forward, so be prepared to move back, out of the way.
It's very important to not allow excessive stretching of the neck because it can cause injury.

I let him relax a minute, then ask again.

He responds very nicely.

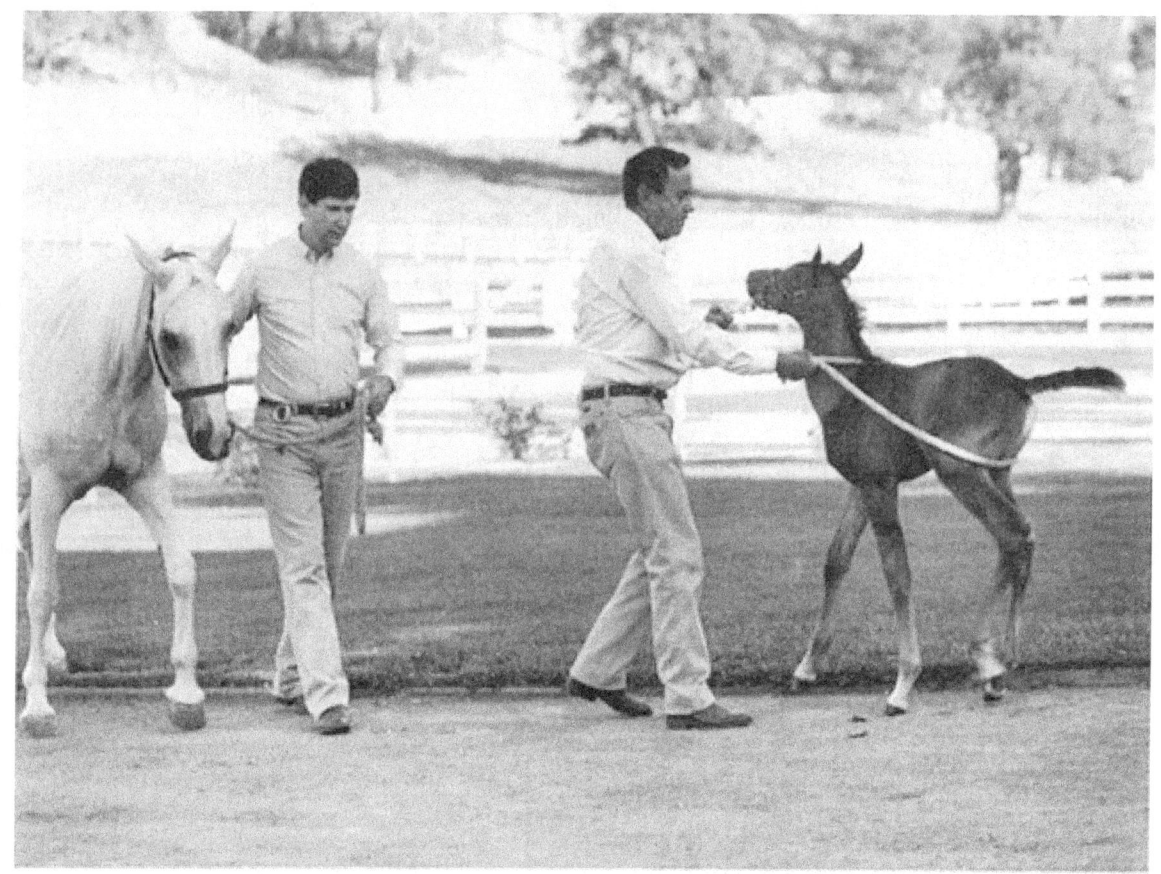

Now I put a butt rope on him. He is already conditioned to move forward when he feels butt pressure.

By letting him follow the mare, he catches on quickly to leading.

Next, I spend a few minutes reinforcing earlier training. He readily yields a hind foot.

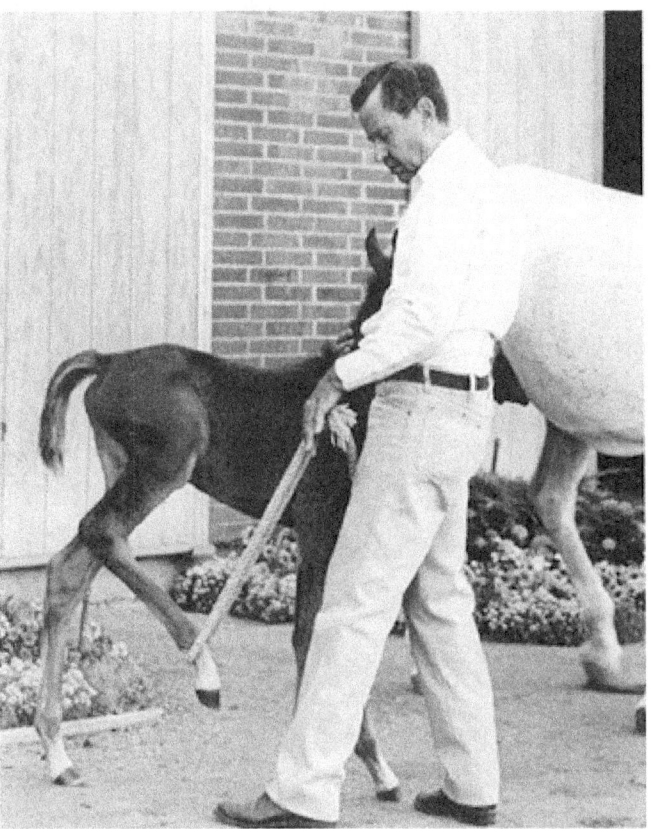
He doesn't object to the foot being held by a rope.

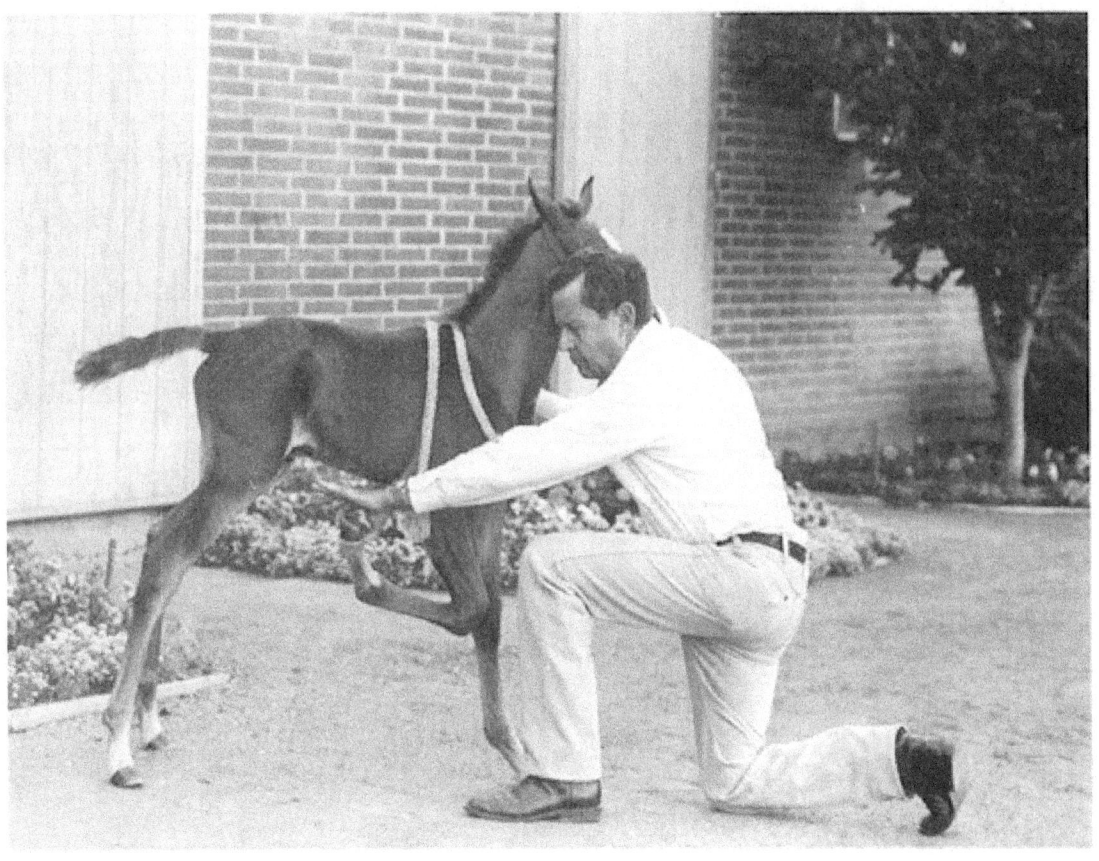
He lets me hold a front foot with my finger tips.

IMPRINT TRAINING

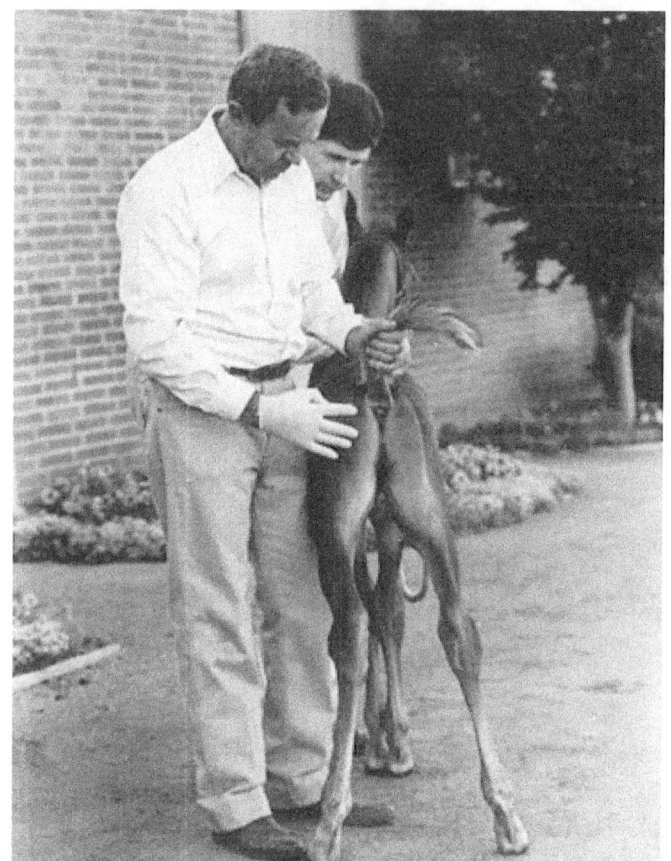

I am reinforcing desensitization of the perineum.

Using a gloved hand and lubricant, I test desensitization of the anus and rectum. He is indifferent, and should always be receptive to a thermometer, or palpation.

He doesn't even object to a rope under his tail.

These responses will be retained and enhanced if we periodically repeat and reinforce the procedures.

10 Teaching The Foal To Tie

Horses can be taught to lead and tie at any age, even at 1 day of age.

THE FOAL should be taught to tie as early as possible; but first, a word of warning.

Tying horses of any age that have not been adequately trained to lead is one of the more common causes of injury. When an untrained horse finds himself tied, his immediate reaction is to pull back and try to run away, since flight is the normal survival behavior in this species.

Unable to get away, the horse will make violent attempts to free himself, and the results can be disastrous. If the halter or the tie rope breaks, the animal will often flip over backwards; this commonly causes fractures of the skull, neck, withers, or back. If he "sits down" violently or slams his hindquarters against a hard object, this can fracture the seat bones of the pelvis. Bruising of the hocks will often result in capped hocks.

Injuries to the skull or spinal column can be fatal, or they can produce central nervous system damage which can be permanent. Many horses are left neurological

Tying lessons learned in foalhood will last a lifetime.

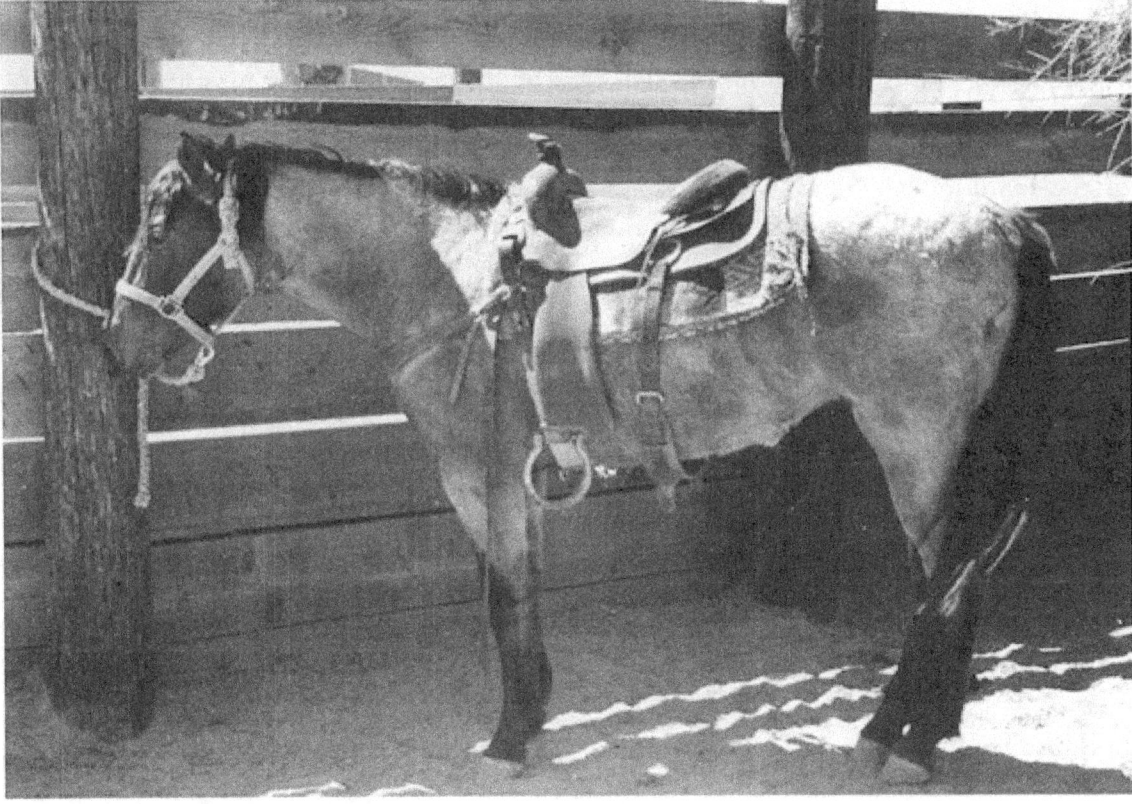

Before learning to tie, the foal must be leading in a straight line. When he feels pressure from the lead rope and halter, he must learn to step forward. This reaction can be reinforced with a butt rope, as Mike Kevil demonstrates here.

cripples by such accidents.

If the equipment doesn't break, the violent attempts to escape may damage the horse's neck permanently. This is one cause of the condition known as "wobblers" wherein damage to the spinal cord in the neck region causes uncoordinated movement, often for life. If the horse is tied too low when he struggles, he can "pull his neck down." This is a serious, permanent injury.

Sometimes the object to which the horse is tied will break. Many horses have been crippled or killed when a tree limb, fence rail, post, board, or panel broke loose and the horse stampeded, "chased" by the thing to which it was tied.

If the horse is tied too low or if the tie rope is too long, a foreleg can get hung up in the rope. This can cause rope burns, broken legs, or spills which might result in a broken neck. Tie ropes, especially if made of nylon, can break and snap back like elastic. I have seen both horse and human eyes lost in this way. Therefore, when tying horses, the following rules should always be observed:

1. Don't tie a horse of any age unless he has been trained to lead and not pull back.
2. Whenever a horse is tied, be sure that the halter, the tie rope, and the snap (if one is used) are unbreakable. Snaps made of cheap, brittle metal are notorious for causing injuries when they break. Be sure that snaps are strong and massive. Ideally, they should be milled of solid brass or stainless steel. Cheap snaps can be deadly.
3. Always tie horses with a slip knot so that in case of an emergency, they can be

If the horse is tied too low, he can "pull his neck down."

TEACHING THE FOAL TO TIE

This is a 2-day-old Quarter Horse/Thoroughbred filly learning to lead. She has already learned to step sideways in response to a light pull on the halter. I demonstrated this with another foal in a previous chapter. Here, I am reinforcing that lesson with this filly.

easily released. Always carry a sharp pocket knife when around horses. A sharp knife can save a human or a horse's life.

4. Always tie horses short and high. If the horse can reach the ground to graze or nibble, he is tied dangerously low, or too long. Ideally, the tie point should be as high or higher than the withers. Then, if the horse does pull back, serious neck injuries are less likely than if he is tied low.

Of course, horses can be taught to lead and tie at any age. The advantages of doing it early, at 1 to 3 days of age, are several:

1. The foal will learn much more quickly, because we will be training him during the best learning period in his life.

2. The foal will be weaker and therefore easier to control than he will be later.

3. If a mishap occurs, the baby foal is much less likely to injure himself now, when he is light in weight and flexible, than later on when he is heavier and stronger. This is especially true if the training procedure is done on a soft surface. If the

Now I want the filly to learn to step straight ahead when I give a light pull. She is responding well.

ground is not soft, it can be deeply bedded. Straw is slippery, however, and is therefore not the best bedding to use; woods shavings are preferable.

Wood shavings would be better.

Before learning to tie, the foal must be leading in a straight line. Do not tie him if he will only follow the lead rope to one side or the other, or lead in a small circle. He must lead straight ahead, even if he does so haltingly. To put it in technical terms, the foal must be conditioned to move forward at least one step when he feels pressure from the lead rope and halter.

This reaction can be reinforced with a butt rope. Always ask for forward movement by pulling on the lead rope first.

The butt rope should only be used if there is not a prompt response to the lead rope. It should not be used as the primary signal, nor should it be used routinely. Once again, the response to the butt rope depends upon prior conditioning to hand pressure, as described in an earlier chapter.

I move back a few steps, and ask her to lead toward me, and she does it readily.

To teach the foal to tie, I use my body as a fence post, and hold the lead in my hands. Facing the foal, I very slowly lean backwards until all the slack is gone from the lead rope, and the foal feels the forward

Once aiain, I move back a Jew steps, and ask her to lead straight ahead. For a few seconds, she didn't respond, then she hopped forward. That's okay. She has already learned that anytime the lead rope is taut, moving toward it will relieve the pressure, whereas flight away from it will not.

TEACHING THE FOAL TO TIE

For the next step, I loop a long lead rope through an inner tube, and position myself behind the filly. I slowly pull the lead rope until it is taut.

As the filly feels the pressure from the lead rope, she steps forward—and is instantly rewarded by slack. Because this filly stepped forward in response to such a light pull, it's difficult to see the difference between tautness and slack in the lead rope. My presence behind her encourages her to step forward—and I can also help her immediately if she panics.

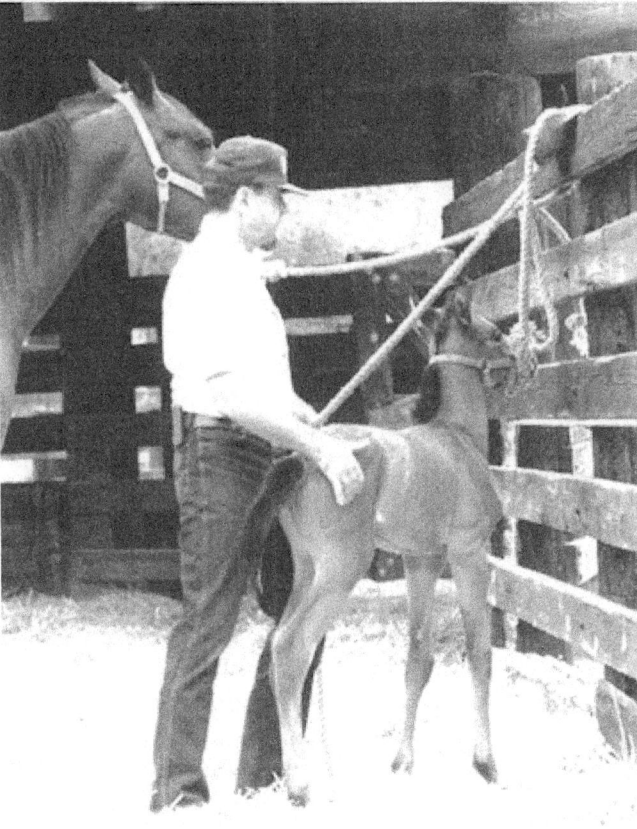

A couple of more steps and she's right up to the fence, which ideally should be a solid wall. If the filly were to get upset, she could stick a front leg through the boards. Note that the mare is tied right beside the filly, so neither one gets upset about being away from the other.

pressure. Since he has been previously conditioned to move forward anytime he feels such pressure, he will automatically move toward me. As soon as he does, I immediately move my hands forward toward the foal to relieve the pressure. I repeat this several times.

What the foal is learning is that anytime the tie rope is taut, moving toward it will relieve the pressure, whereas flight away from it will not.

As soon as this response is automatic, we are ready for the next step. For me, this is usually just a few minutes' work, but there is no hurry. There is no reason not to stop and continue the lesson at another time if things are not going well and the foal fatigues or loses attention.

The next step is to fix an automobile inner tube to a fence or post or tree limb. A solid wall is ideal, so that if the foal rears, he won't stick a leg through the fence. The rubber inner tube should be higher than the foal's head. Now loop a long lead rope through the inner tube and position yourself behind the foal. Slowly pull the lead rope until it is taut.

As the foal's neck starts to extend, the inner tube will begin to stretch. The least movement forward will be instantly rewarded by the release of pressure. The inner tube is duplicating what you did with your arms when you served as a fence post. Immediate reward! Positive reinforcement! In case the foal panics and tries to go backward, your position directly in back of him will inhibit this movement and will protect the foal from flipping over backwards.

Repeat this lesson daily until you are quite sure that the foal will move forward to relieve the pressure, not backward. It helps if the mare is tied alongside the foal. It reassures the foal and lessens the desire to flee, because he feels secure next to the mare.

When you are absolutely sure that the foal will not pull back, you are ready to tie the foal and leave him for a minute or two,

Horses that are correctly broke to stand tied will stand quietly when tied to anything—even overhead picket lines

but don't actually tie him. Instead loop the tie rope around the inner tube twice. Then stand a few feet behind the foal until, as he moves about, the tie rope is drawn taut. If the foal immediately moves forward, the job is done. The foal is taught to tie. But, be prudent; do not tie the foal with an actual knot (not even a slip knot), and do not leave the foal alone until this lesson has been repeated many times. Be sure that the mare is always close to the baby foal, so that neither will panic.

11 More Halter Training

It is not reasonable to expect a baby foal to stand tied for any length of time.

ONCE THE foal is leading well and will stand tied, he is ready for some advanced halter-training techniques. But first, let's define our terms:

Leading Well. With the baby foal, this means following the lead rope in any direction, but at a slow pace. The foal should not yet be expected to lead too far away from the mare without showing resistance and anxiety. Nor should he be expected to lead any faster than at the walk. Therefore, do not ask him to do those things. If the foal compliantly follows you at a walk in various directions close to the mare, he is doing very well.

Standing Tied. This means that the foal will stand for a minute or so when tied to a post or fence, short and high, near the mare. It is unreasonable to expect a baby foal, bonded to the mare and imprinted to follow the mare, to stand quietly while tied if the mare is any distance away. Nor is it reasonable to expect a foal of this age to stand tied for any length of time. A foal quickly tires of standing and will want to nurse, to move about, or lie down. Therefore, do not ask the foal to do what he cannot reasonably be expected to do.

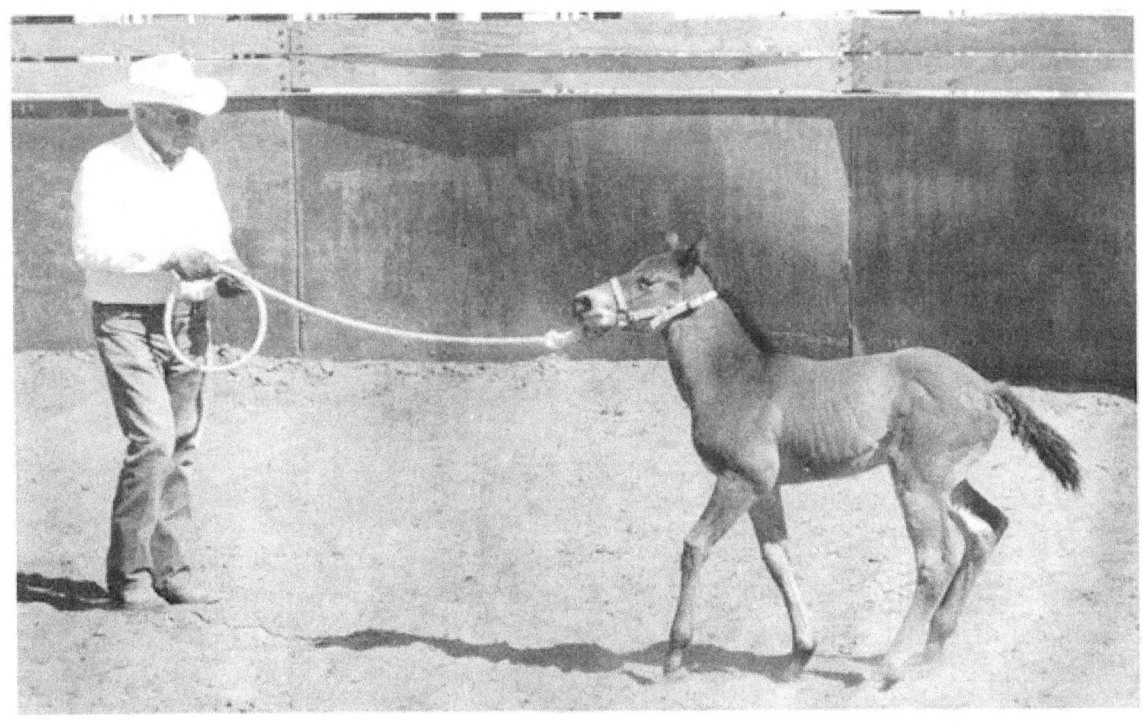

This baby foal, being handled by Jack Brainard, Aubrey, Tex., readily steps forward in response to a very light pull on the lead rope

She is an excellent example of a foal who leads very nicely.

At this stage, the foal may pull back against the rope at times, but rather than panicking, attempting to flee, or throwing himself, he will either bounce forward to put slack in the rope or gently lean forward to relieve the tautness of the rope.

If the foal meets the above criteria, and this can be seen as early as 2 or 3 days of age if everything has been done correctly and you are dealing with a cooperative foal, advanced halter-training techniques can be started, such as:

1. Leading from the mare.
2. Leading in hand at various gaits.
3. Standing quietly in hand.

We will discuss the first technique in this chapter, and the other two techniques in subsequent chapters.

Leading From the Mare

It is very beneficial to lead the foal from the mare if she is well-mannered and well-disciplined. The mare does not have to be sound of gait, because we are going

It is very beneficial to lead the foal from the mare.

I usually tie my horses and mules to an inner tube, using an unbreakable halter and lead rope, until they are 2 years old. This yearling mule was imprint trained, never pulls back, and leads easily.

Horses cannot apply what they learn on one side of their body to the opposite side. That's why the learning process must be repeated and duplicated on each side.

to move slowly around a ring or corral to start with. Mount the mare and hold the foal's lead rope in your right hand with the foal on the off side of the mare. Do not secure the lead rope in any way to the saddle, or dally it around the saddle horn. Slowly move the mare a step or two, and then stop and allow the foal to do the same. If necessary, a loose butt rope can be used to encourage the foal forward.

When the foal steps forward, move the mare ahead another step or two. Progress until the foal leads willingly. Do not drag the foal. This only teaches the foal to drag, not to lead. Soon the foal will be leading alongside the mare at a walk around the perimeter of the ring. Then switch the foal to the near side of the mare. Remember that horses cannot apply what they learn on one side of their body to the opposite side. That's why the learning process must be repeated and duplicated on each side. If, at the end of this session, the foal leads obediently around the ring alongside the mare in either direction from either side, be satisfied.

In a day or so, repeat the lesson and start to add a few turns in the opposite direction. When the foal has mastered that, try some figure eights. Remember that all of these leading exercises should be done from both sides.

At intervals, when the foal performs well, lean down from the saddle and stroke his head, neck, and body. This has two purposes. First, it serves as a reward to positively reinforce the satisfactory behavior. Second, it accustoms the foal to seeing and feeling a person above him. When the foal has grown up and is mounted for the first time, he will be much less likely to panic when he sees and feels the rider above him. In fact, I like to straddle the newborn foal for the same reason shortly after he first stands. Obviously, no weight should be put upon his back if this is done. Some foals are also too tall for some people to do this.

If the mare is sound enough to trot, that should be the next phase of training. All of this can be taught from the ground, as we'll describe later. But, if he is taught from the mare, the foal will learn more quickly because he wants to stay alongside the mare.

Once the foal leads well at the trot, try cantering slowly for a few yards. Do not rush things. We don't want the foal to get excited, pull back, fall, or escape. That is why this should be done in a ring or corral. The foal can then initially be kept between the fence and the mare.

The Fixed Butt Rope

After the foal has been led from the mare several times and will do so at the walk, trot, and canter, I switch from a loose butt rope (which is simply a loop around the hindquarters, separately controlled from the lead rope) to what I call a fixed butt rope. This step in halter training is not necessary, but, if done correctly, it can teach a week-old foal to travel in collection.

How does it do this? In order to understand, we must first explain what is meant by collection. The rider must consider the horse as having two moving parts: the forehand or front half and the hindquarters or back half. Since the hindquarters are the source of propulsion, they are in the best position to do this if they are gathered under the horse, rather than stretched out behind.

The skilled rider brings the hindquarters forward under the horse with pressure from his legs and the seat. Simultaneously, forward motion is inhibited by the hands through the use of the bit and reins. Thus, the collected horse at any gait has a rounded back; the hindquarters are under the horse; and the head is flexed at the poll to bring the face into a more vertical position. The position of a collected horse is similar to the position of a human football line-

IMPRINT TRAINING

Once a foal leads well from the ground, I like to lead him from the mare. This series of three pictures shows how I use a loose butt rope, which is actually an old catch (lariat) rope.

I keep the foal's lead rope in my left hand, and the coils of the catch rope in my right hand. Note that neither rope is tied or dallied to the saddle; that could be dangerous. I can also quickly drop the coils of the catch rope, if necessary, to avoid getting my hand entangled.

Because this filly was imprint trained and sensitized to butt pressure at 1 day of age, she readily moves forward when she feels pressure from the rope.

man. The spine is slightly flexed, the legs are flexed under for maximum pro pulsive power, and the head is slightly flexed. In this position the horse is best able to run, jump, change directions, whirl, and stop.

Collection does not mean that the head is carried low with the nose towards the ground. In this position the weight is on the forehand; the horse seems to be going downhill. This position puts more of the horse's weight on its non-propulsive front legs.

Nor does collection mean that the head is so flexed that the face is carried behind the vertical. Excessive flexion of this kind, often seen in the show ring, inhibits the athletic performance of the horse and the control that the rider has over the forehand.

Assuming that the foal has been correctly conditioned to lead, instead of pulling back, and is conditioned to move forward when he feels pressure on the butt area, he will, in just a few minutes, learn to move at all gaits in collection and with impulsion.

The fixed butt rope is a rope separate from the lead rope, and which runs from the bottom of the halter around the butt back to the halter. It must be exactly the right length. If it is too tight, it will inhibit the action of the hind legs to fully extend backwards. If too loose, it will achieve nothing. Study the photographs. When the foal is standing in a normal, alert, but relaxed position, the fixed butt rope lightly contacts the hindquarters. It also allows the foal to carry his head in a normal position, very slightly flexed at the poll.

As the foal is led, if he pulls back and elevates his nose, he tightens the fixed butt rope, bumping his hindquarters and his nose. To avoid these pressures, the foal will bring his hindquarters under him a little, and flex his head slightly. Immediately he has assumed the correct position for collection.

Even if he attempts to lag behind, in this position, the butt rope applies pressure, urging him forward. Leading the foal with the fixed butt rope quickly teaches him to move forward eagerly to stay alongside of the mare, but to do so with his head slightly flexed, his back slightly rounded, and his hindquarters slightly under him.

Remember, the butt rope must not be too tight. If it's too loose, the only harm is that it will fail to teach impulsion with collection. But, if too tight it will inhibit impulsion because the hind legs cannot fully extend, and forward propulsion is interfered with.

Start with the fixed butt rope rather loosely rigged. Later, it can be tightened up a bit at a time. Once the foal is leading responsively with the head slightly flexed and the hindquarters slightly under his body, do not tighten the fixed butt rope any more.

In order to understand the importance of collection, it must be realized that this is not a normal body position for the horse at liberty. The horse in a flight position elevates the head and neck with the nose extended. Tucking the nose during flight is not natural to the horse, but it is a necessary position in order for the rider to communicate with the horse via the hands, the rein, and the bit, and in order for us to control the forehand. It is not necessary to be on the horse's back to teach collection. It is often done from the ground by various means, including bitting rigs, driving, and longeing. With the fixed butt rope, a very young foal can be taught early in life that free forward motion is possible in a collected position.

If the mare is sound enough and of suitable disposition, it will be beneficial to lead the foal from her outside the training ring. Alternately lead from both sides of the mare. If the mare calmly accepts highway traffic, livestock, tractors, and other frightening objects, the foal will be reassured, and he will learn to accept them, too. While riding the mare, lead the foal through running streams, choosing shallow water initially. Lead over bridges, through gates, and past cattle, dogs, and anything else the foal will be exposed to during his lifetime.

If the mare is not sound enough or well trained enough for the described leading procedures, they can all be taught from the ground as described in the next chapter.

Then, after weaning, the foal can be led from any gentle, well-mannered horse.

While being led from the mare or other lead horse, most foals will try to nibble at the rider's leg, the lead horse, or the stirrup. This should be sharply discouraged by *immediately* bumping the foal's nose with the leg. Usually after several such bumps, the foal will be discouraged from nibbling. As in all training procedures, the swifter the punishment and the more subtle it is, the more effectively it will discourage repetition of the behavior.

Once the filly leads well with a loose butt rope, I put a fixed butt rope on her. It is snapped to the halter, goes around her hindquarters and back to the halter ring, where it is tied off. I then use the end of the rope as the lead rope. When I pull on it, it does put pressure on the halter, but not on the butt rope. However, when the foal raises her head in response to a tug on the lead rope, that does put pressure on the butt rope. The butt rope should be snug enough to keep the hindquarters under the foal, and the nose tucked in a bit; but it should not be so tight that it inhibits full backward extension of the hind legs.

When a foal lags behind when being led, he normally tries to raise his head. When he does this, the fixed butt rope applies pressure, urging him forward. To avoid causing this pressure, the foal quickly learns to travel with his nose down and his hindquarters under him, which is the basis for collection.

12 Advanced Halter Training

RELATIVELY FEW suckling foals of any breed are shown in halter classes at horse shows, but it is a good idea to teach every foal good halter manners. They are especially necessary if a foal is shown. As with all other techniques in this book, good manners can be taught to foals at just a few days of age.

Once the foal is leading well in hand as previously described, we can refine the techniques. We have already taught the foal not to pull back or lag behind when led. The pressure of the halter at the back of the neck discourages pulling back. At this stage of the game, therefore, a halter made of strong thin cord is more effective than one made of flat, broad material, because it applies more discomfort if the foal does pull back. If necessary, pulling back can be further discouraged by a rein forcing, loose butt rope, especially in the early training stages. I have not found it necessary or advisable to use a war bridle type of halter to teach foals to lead, so I do not recommend one.

We must also teach the foal not to pull ahead when being led, and there are a variety of techniques used to accomplish this. All of them involve the same principle: if the foal moves too fast-ahead of the person leading him—he is subjected to an uncomfortable and intimidating experience. This discourages the foal from pulling ahead, and soon he will stay at the leader's side, just as a dog learns to heel.

Foals should be taught how to lead well and stand properly-especially if they will be shown. This is Larry Myerscough of Houston with a weanling.

However, dogs are usually taught to heel at the handler's left side, whereas horses are taught to lead on the handler's right side. Ideally, both species should be taught to heel or lead on either side.

The photographs show three methods

ADVANCED HALTER TRAINING

Here's how I like to lead a halter-broke baby foal. My goal is to teach the foal to follow my right hand. I am also holding a riding crop in my right hand. If the foal starts moving too fast, I tap him on the chest with the crop. If he moves too slow, my body will tighten the lead rope, urging the foal forward.

Here's another way I often lead a baby foal. I take a half-wrap with the lead rope around my elbow and forearm. If the foal moves too fast, a quick jerk backwards with my elbow will snatch the foal and slow him down. If he moves too slow, a jerk forward will urge him forward. In either case, the foal learns to follow my right hand, and will keep a slight amount of slack in the lead.

of teaching a foal to lead correctly without lunging ahead of or crossing in front of you.

When a horse is full grown and he persistently gets ahead of the person leading him, or crosses in front of that person, it not only makes walking difficult, but it sends a message to the horse that he is in control of the situation. It creates disrespect. It is not conducive to an attitude of submissiveness on the part of the horse. It means that the person leading cannot control the horse. It is also a good way to get your toes stepped on if the horse crosses in front of you; or worse, if the horse pulls in front of you, he could kick and injure you.

Next, we must teach the foal to halt when we halt. The simplest way to do this is to stop abruptly, tightening the lead rope, and them immediately releasing the pressure when the foal does stop. I like to combine this with the command "whoa" which I pronounce as a sharp, but not a loud "ho!" But it isn't necessary to use a verbal command, and some trainers prefer not to, especially since many riders misuse the command, saying whoa when they merely want the horse to slow down. Used in that way, the word soon becomes meaningless to the horse.

Verbal commands, like all other command signals, should call for a single response and should not be varied.

Whoa means stop, and it means an abrupt and total stop. It should not mean slow down or calm down, or stop for a moment. It means "stop immediately and completely and you are to stay stopped until you receive another command" As the foal stops, turn toward him. Facing him will block his progress, teaching him to stand still, and to not resume forward movement prematurely.

The foal is rewarded for stopping by the relief of halter pressure, and this can be reinforced by briefly rubbing the neck or face and an expression of praise such as "good boy" or "good girl" or simply "good" ut-

In this method, my hand is in front of the foal, but it does not hold the lead rope. Instead, the lead rope is held in the left hand, with the rope passing in front of my waist. If the foal moves too fast, I can snap my right hand downward to jerk the lead rope and discourage rapid movement. If the foal moves too slow, the lead rope will tighten across my waist and pull the foal forward. The foal learns to stay just behind my fist.

tered in a soft, soothing voice.

Be consistent in your commands and in your reinforcement techniques, whether positive or negative. Saying "good boy," one time and "what a smart fellow," the next time and "you did it right," the third time confuses the horse.

Similarly, a horse of any age can quickly learn that "bad!" is an expression of disapproval, especially if said sharply.

As the foal learns to stop when you stop, you can gradually eliminate turning toward the foal to block his progress.

Your hand holding the halter rope will become the focus of his attention and when your hand stops, blocking his path, the horse will stop. The verbal command, if you choose to use one, should be as consistent and as sharply defined as the visual command.

With the foal now leading and stopping correctly, our next goal should be to teach him to back. When he was only a few hours of age, we taught him to back away from the pressure of our finger tips at the base of the neck or the upper part of his chest. All we need now is to face the foal, apply pressure to his chest, move our body towards the foal, and he should step backward at once. This movement is rewarded, as always, with immediate relief of pressure. We pause for 10 or 15 seconds, and repeat the process, this time combining it with backward pressure on the lead rope.

If we want to, we can simultaneously and sharply say, "Back!" If the foal takes one step back, immediately reward him by releasing the pressure and reinforce positively with praise and rubbing.

Very quickly the foal should step back on command, and pressure on the base of the neck or chest can be gradually eliminated. However, that signal is in the foal's memory, and the request to back can be reinforced with pressure in that area.

Progressively, as a second step back, and then a third, occurs, more can be requested. The foal should then lead at the walk obediently and on a slack lead rope. He should stop when commanded to. He should back on command until signaled to stop.

ADVANCED HALTER TRAINING

This is an older colt trotting alongside the handler in this photo . . .

. . . but pulling ahead here. Foals should be taught at an early age not to pull ahead

IMPRINT TRAINING

Every foal (and horse) should be taught to lead with the handler on the off (right) side, but this is NOT the way to do it.

One way to do it: carry a long-enough whip in one hand; cluck to the colt to ask him to move forward; then tap him on the butt with the whip. Ideally, the foal should be positioned with a fence on his left side so he will move forward-not sideways- in response to the whip.

ADVANCED HALTER TRAINING

Pretty quick he will lead just as well from the right side as the left.

Learning to stand quietly and at attention. Start by requesting it for only 5 or 10 seconds. Praise him for a good response.

The foal can also learn to position his feet correctly for being shown at halter.

Teach the foal not to pull ahead when being led.

Two photos of an older colt leading well at the trot.

This colt is being taught to back. The handler is moving his body toward the colt, applying pressure to the colt's chest, and saying, "Back!" As shown earlier in this book, backing can be quickly taught at 1 or 2 days of age.

Standing In Hand

While being taught to lead correctly, the foal should also be taught to quietly stand still at attention. When teaching this, start by requesting it for only brief periods of time, perhaps only 5 or 10 seconds.

Praise a good response. Gradually lengthen the time periods the foal is asked to stand quietly at attention.

It is helpful to do this close to where the mare is tied, so that the foal does not lose attention due to anxiety. If the flies are bothersome, an insect repellent should reduce the foal's fidgeting.

The week-old foal can also learn to position his feet correctly for being shown at halter. One need not wait until weaning time to teach this. It can be taught to foals under a week of age, and they will remember it later on, because the foal was taught as a newborn to move forward and backward in response to pressure. Since he learned to move either the forehand or the hindquarters laterally in response to pressure, it will be easy to slightly alter his stance or leg position once he learns to stand quietly when asked to.

Thus far, we have taught all of the in hand maneuvers at the walk. We can now teach the foal to move at faster gaits. If the foal was taught to lead from the mare, he already knows how to trot to keep up with the mother. Now that we are leading him from the ground, we can take advantage of a quality know as allelomimetic behavior, or mimicry. Foals will imitate what the mare does. If she is spooky, the foal will shy at what she shys at. If she boldly crosses a stream, the foal will want to do the same. If she paws or cribs, there is a good chance that the foal will learn the same vices.

Foals can be taught to move forward when led at faster gaits with the use of pain-inflicting devices such as war bridles or whips, but this shouldn't be necessary. If the foal has been imprinted to follow, trust, and be bonded to a human, he will be inclined to mimic human behavior.

So, if when leading the foal, we start to jog in place, then gradually increase our forward speed, the foal will usually respond by doing the same. Reward the first brief response with extravagant praise, and don't be in a hurry to repeat it right away. Stopping a lesson is in itself a reward to a horse.

This colt is being taught to back. The handler is moving his body toward the colt, applying pressure to the colt's chest, and saying, "Back!" As shown earlier in this book, backing can be quickly taught at 1 or 2 days of age.

When doing this, don't look back at the foal. If you turn toward the foal, it will discourage forward movement. This is true of horses of any age. It intimidates the horse if the person leading him looks at him. His natural response, if he respects the person, is to maintain a certain distance and to respect the other individual's personal space. Therefore, forward motion is inhibited. So, if you *are* teaching a foal to lead at a trot, assuming that he is already well trained to lead at a walk, simply look forward. Then start to trot (in an exaggerated manner) in place, and tighten up the lead rope. Don't look back! You may even let the lead rope lengthen a foot or so. (Be sure to use long lead ropes when training—10 or 12 feet is not too long.)

The foal will feel that you *are* moving away, and eventually should pick up his gait and start to trot. When he does, slowly (not abruptly), decrease your speed, and as the foal comes to you, quietly reward him with praise and rubbing. An occasional food reward, such as a handful of grain, is effective in training horses, but is not necessary. Horses, being animals that live in groups—as do dogs, humans,

Ponying foals is a good way to improve their leading.

dolphins, chickens, pigeons, baboons, whales, antelope, and many other species—crave acceptance. That's why praise is so effective in reinforcing behavior in group species, whereas loner species, such as cats and bears, do not respond as well to praise as they do to food rewards.

ADVANCED HALTER TRAINING

Ponying also accustoms a colt to a rider above him ... and to the rider handling his ears and rubbing his back. All of this will pay off when the colt is old enough to be broke.

13 Teaching Perfomance Basics to the Young Foal

SINCE THE critical learning periods in precocious species such as the horse are in the immediate postpartum period, it follows that certain elements of performance training can be effectively taught during the first week of life. Moreover, these things can be taught in less time and with better retention than at any other time in life.

Prospective jumpers—In addition to everything we have previously mentioned, very young foals can be taught to jump fences in just a few minutes. Using poles, boards, or rails just a few inches above the ground, divide the ring or corral into two sections. Ride or lead the mare over the dividing barrier to the opposite end of the enclosure. The loose foal, separated from the mare, will read with anxiety. He might initially balk at the dividing "fence" but eventually will jump it. Allow the foal to rejoin the mare, and then repeat the lesson in the other direction. Keep it up until the foal, easily and without hesitation, jumps the barrier, then quit for the day.

A day or so later, repeat the lesson, raising the rails an inch or so. After a few lessons, the foal will easily jump an 18-inch barrier. Don't discourage him with a barrier that is too high. (additional height might also cause too much concussion and stress on immature bones, tendons, etc.) At this point you are not looking for height.

Straddling a young foal and swinging a rope over his head will habituate him to this stimulus.

TEACHING PERFOMANCE BASICS TO THE YOUNG FOAL

Young foals will actually bond with calves if they are kept together.

When the foal is older, he will have no fear of a rope, as these two photos show.

You want the foal to develop confidence and form, so make it easy. After he easily jumps 18 inches, substitute a log for the fence. The foal will have learned the basics of jumping that he will forget.

Prospective western horses—Let the foal see cattle. The mare and foal can be put in a pen adjacent to some cattle. Better, they can be put in a pen together with some cattle, but be sure the cattle are gentle and harmless. A halter-broke calf can even be tied next to the foal. Young foals will actually bond with calves if they are kept together. You want the foal to learn not to fear cattle, and to be used to their smell and appearance.

Standing beside, or actually straddling the young halter-broke foal, swing a rope over his head until he is habituated to this stimulus. If the foal has had the imprint training procedures described in the earlier chapters of this book, it should take less than a minute to desensitize the foal to a lariat whirling over his head, and the rope being thrown and then dragged toward him.

I once taught a 3-day-old foal to "work" a rope, keeping it taut when it was attached to the saddle horn of a tiny pony saddle.

Driving—If the foal has been taught to lead as described earlier, a line can be attached to each of the rings on the cheek pieces of the halter and the foal driven forward while you walk behind him. It isn't necessary to do this, of course, but it emphasizes the fact that foals can be taught anything at any age, and the earlier he is taught, the less time it takes and the better it is retained.

Dressage—A lot of basic dressage prin-

111

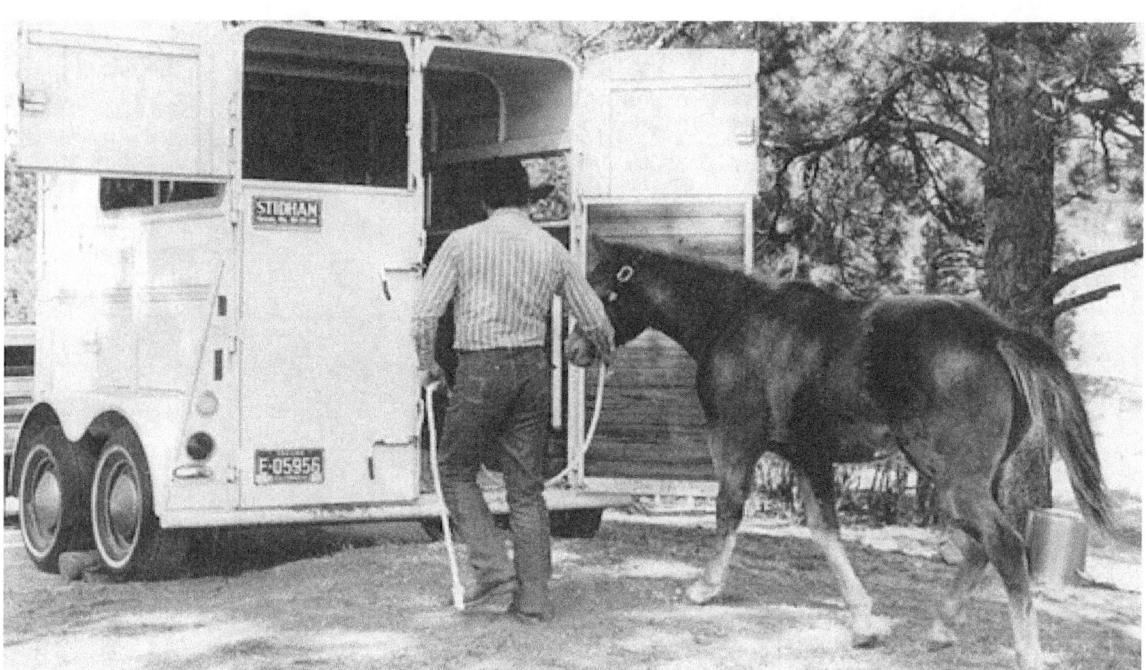

Sooner or later, almost every modern horse needs to learn to trailer. Why not teach it to the baby foal?

Horses can be trained to load into a trailer and travel quietly at any age. And the sooner it's done, the better. This 11-month-old colt has graduated from Trailer Loading 101, and walks in with no hesitation.

ciples are involved in the training procedures already described. A week-old foal is on his way to becoming a dressage horse if he leads lightly in any direction at any gait, stands quietly and backs up, rotates on his forehand or hindquarters in response to gentle pressure, and moves out in collection with his hindquarters under him and his head flexed at the poll.

Beyond that, I am certain that advance dressage movements can be taught to young foals, although I have never tried it because I lack the expertise to do so. However, legendary California trainer Jimmy Williams teaches colts the piaffe on a treadmill before they are ridden. I have no doubts that the same thing can be done with a 2-week-old foal, but I question if there would be any practical value in it.

Certainly, longeing techniques can be taught to young foals. I have even taught mules just a few weeks of age to longe. Be aware, however, that excessive longeing is hard on young legs.

All horses—All horses will benefit from all of the above-mentioned training, whether their ultimate use will be as English, western, dressage, or driving horses. Versatility is a valued attribute in any horse.

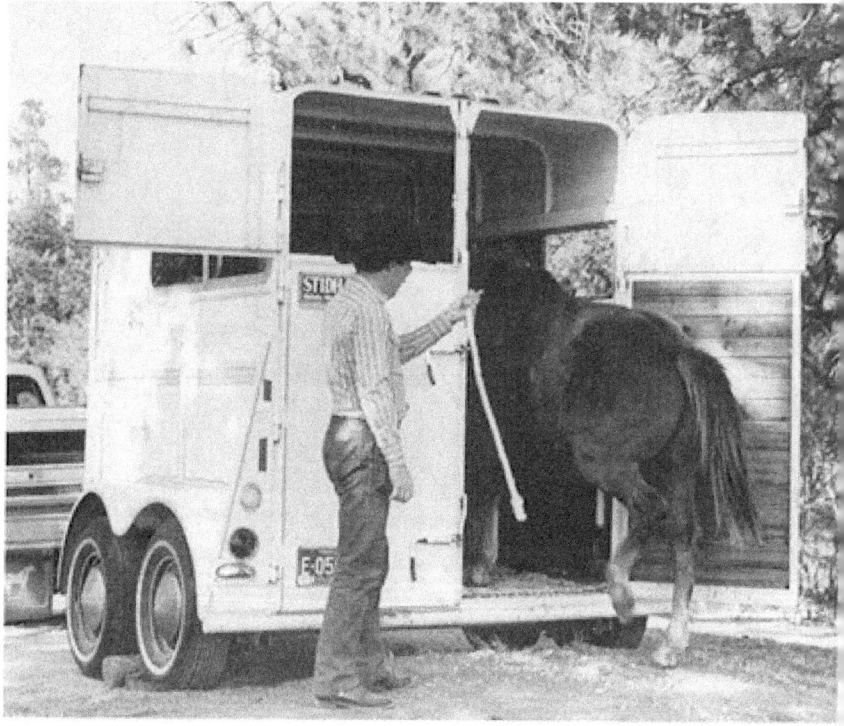

One thing that every horse should do is to trailer quietly. Possibly nothing else causes horse owners more grief and frustration, and injures and kills more horses, than does trailering. It causes people to lose their tempers, which results in horses being abused, which ruins their attitudes towards

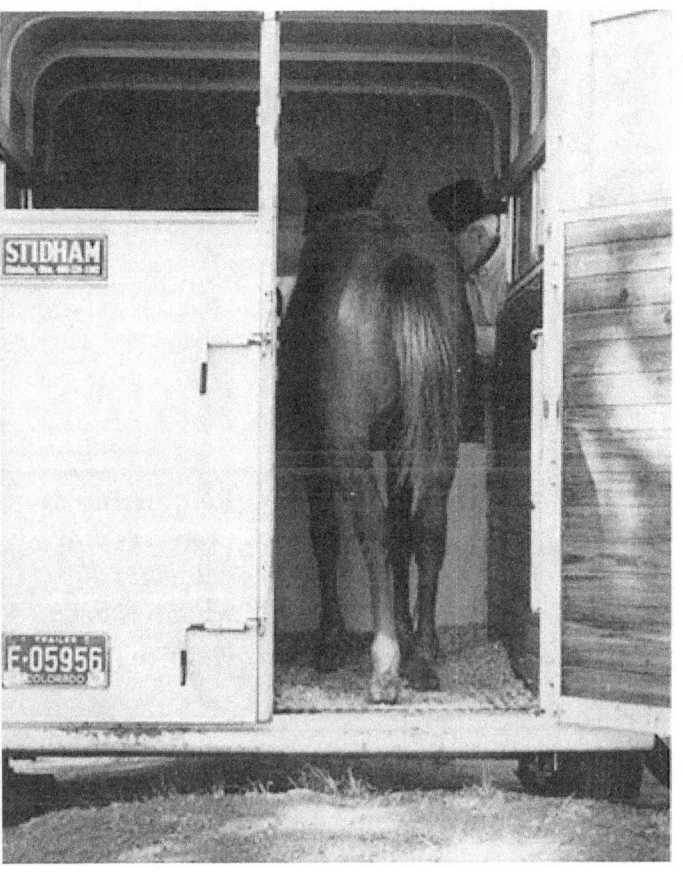 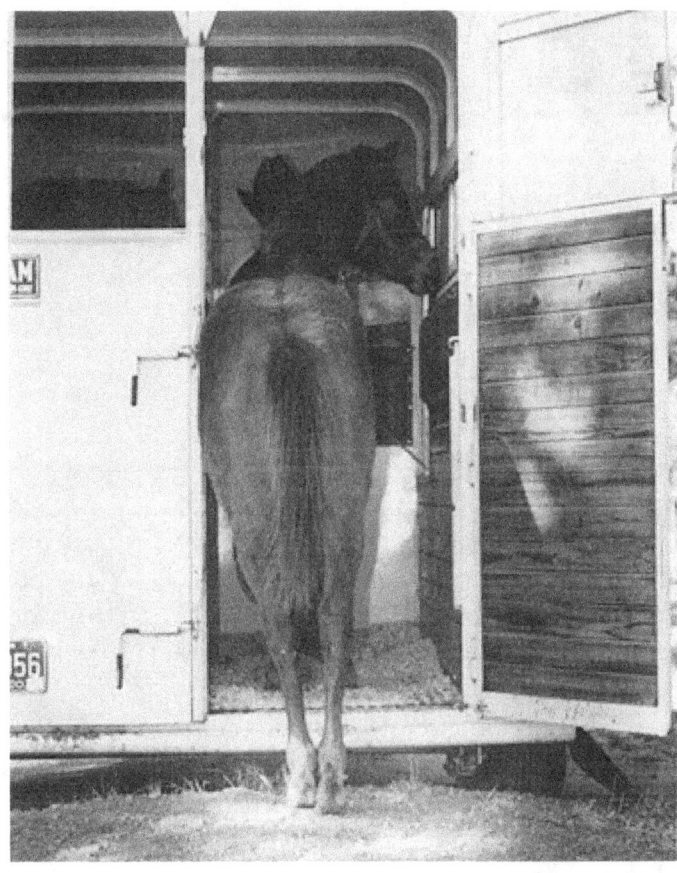

trailering.

Horses can be trained to load into a horse trailer and travel quietly at any age. (even if the horse already hates and fears trailers, that behavior can be modified by a competent horseman.) But, why wait? Sooner or later, almost every horse needs to learn to trailer. Why not teach it to the baby foal? Put the mare in one side of the trailer. Put the foal alongside of her.

When they are both relaxed, take them for a short, slow ride. Repeat the lesson three or four times before the foal is *2 weeks* of age. The job is done! The foal has learned not to fear the shaky, noisy, scary enclosure. It has learned that it is in a safe, secure sanctuary. (Note: Foals this young should not be tied in the trailer.)

Alternately, park the trailer in a corral. Either leave it hitched to the vehicle or block it up securely so that it is stable. If it is a step-up trailer, which I think is safer than a ramp-loading trailer, and the step up is too high for the baby foal, dig a hole for the wheels to lower the floor closer to the ground. Securely tie the doors open, and close the escape doors. Put the mare and foal in the corral. The foal should be at least 1 month of age-old enough to be eating solid food.

Put the feed on the trailer floor toward the back of the trailer. Each day, put the feed a few inches farther forward. In 10 or 15 days the mare and foal will be going in and out of the trailer whenever they want something to eat, and will no longer fear it. Although this method works best with a stock trailer, it can also be done with a two-horse side-by-side.

One precaution: If the foal lies down near the trailer and gets a leg under the trailer or under the ramp, he could injure the leg when he attempts to get up, so such space should be blocked off with planks or cinder blocks.

14 Reinforcing Responses

IN THIS book I have repeatedly said that training newborn foals during their early imprinting and critical learning periods is advantageous because of the speed of learning, and because what is learned is so permanent. This is true, but that doesn't mean that periodic reinforcement training sessions should not be done. Reinforcement not only preserves what has been learned, but it also enhances it.

Repeated training sessions need not be done frequently. In fact, I believe that they can actually be overdone. Once all of the procedures described have been done and the foal is a few weeks old, a weekly training session of 15 to 30 minutes should be adequate. In general, we can divide all of the procedures into three categories: bonding, desensitizing, and sensitizing. These will all be discussed separately.

Bonding Procedures

Horses of any age, being group animals by nature, will bond with almost any living thing that they live with, including dogs, cats, goats, children, or adults. If aggression occurs, it is usually because there is not a clear-cut order of dominance. If there is enough space to permit retreat, the submis-

Reinforcing preserves and enhances what has been learned.

Horses that bond with other horses will strengthen that bond with mutual grooming.

Photo by William Roberts

Horses of any age will bond with almost any living thing that they live with. Here are a foal and a goat enjoying some time in the sun shine together.

Photo by Jean Latimer

This foal is getting acquainted with the family dog. Photo by Jana Dougherty

Training newborn foals during their early imprinting and critical learning periods is advantageous.

sive individual will do so, avoiding a physical confrontation with the more dominant individual.

To a horse, a human can serve as a surrogate horse and can therefore become a bonding object. Conversely, horses can become surrogate humans to a person, and this commonly occurs in the horse-human relationship, just as it does when other companion animal species are involved.

When a human imprint-trains a new born foal, a special kind of bonding occurs to the foal and this is the maternal offspring bond. The newborn foal will regard the trainer as he does his own mother, with trust, respect, love, and a desire to follow. This is an ideal attitude for a horse to have, because although trusting, he is also deferential. It is a sub missive attitude, highly conducive to sub sequent training.

With maturity, the nature of this bond changes. Just as the foal grows up and eventually sees his dam as just another horse rather than his mother, and vice versa, the imprint-trained foal ultimately sees his trainer as just another human rather than as his mother. He may be bonded strongly to that person, but it is no longer the maternal-offspring bond. It is, we hope, if all was done correctly, the bond that exists between a herd leader and one of the tractable and subservient members of the herd.

To put it in human terms, the mother and child relationship is replaced by a relationship similar to that between a benevolent master and an obedient, adoring servant. Are they partners? Yes, in a sense. It is a partnership when a human and a horse go through a cross-country course together, or when a human and a horse pen a herd of cattle together. But, let's be honest. One is the master—the boss. Hopefully, it is the human who is the boss.

As training proceeds, the bond between the horse and the trainer strengthens, assuming of course that the trainer does nothing to sever that bond. Trainers who mistreat the horses they work with— trainers who lose their temper, have tantrums, become abusive, or are impatient— do not have the same relationship with a horse that other trainers enjoy . . . specifically, those trainers who are more skilled, and who train with kindness, patience, and respect.

Horses that bond with other horses strengthen that bond with mutual grooming activity. Similarly, the trainer should spend some quiet time with the horse being trained. The horse's withers, neck, or the base of the tail can be rubbed for a minute or two. Quiet country rides should be interspersed with ring work. And, whether in the ring, or out in the country, just standing still and resting together for a few minutes is beneficial.

Desensitizing Procedures

A lot of training done immediately after birth is aimed at desensitization. We have explained that in desensitizing horses to frightening sensory stimuli which ordinarily should provoke a flight reaction, we utilize repetitions of the stimuli in a technique known as *flooding*, in order to obtain a condition known as habituation, wherein

Here's a sequence of photos in which I am reinforcing responses with a mule foal. In this photo, the foal is tied to the fence while I sack her out. She is completely desensitized to the "sack," which is actually a saddle blanket.

the subject is unaware of and non-responsive to the stimuli.

It has been emphasized that when desensitizing, it is extremely important not to stop the stimulus before habituation occurred. To stop when the animal is still showing a desire to escape may fix the escape behavior, and we will then have *sensitized* the horse, instead of *desensitizing* him to the particular stimulus.

Once habituation is obtained, however, there is no need during subsequent reinforcement sessions to repeat the stimulus many times. For example, if we have sensi-

REINFORCING RESPONSES

She is desensitized to the lariat and unafraid of it

tized the ear canal by putting a finger in it and wiggling it 30 or 60 or 100 times until the foal stands placidly ignoring us, we can, a few days or weeks later, reinforce the lesson. We simply repeat the finger in the ear, gently so as not to generate a flight response. Usually the foal seems to say, "Hey! What's this? A finger in my ear? Oh yes! I remember! That was done right after I was born. It doesn't hurt. I'll ignore it"

As soon as the horse's attitude conveys that message, stop the stimulus, praise the horse, and go on to something else. It will, each time, be easier to do, and less likely to excite the horse. The same thing applies to all other desensitizing procedures. At a reinforcing session, it isn't necessary to do everything. Avoid a ritualized approach after the initial training session.

Let's describe a typical session in a month-old foal. We'll assume that he has been fully imprint-trained, but has not been handled for *2* weeks and is in pasture with his dam. Quietly and alone, go out into the pasture and approach the foal. Stop when he is close enough to recognize you. Often the foal will come up to you. Rub his neck and, working slowly, quietly slip the halter on him. Tie the mare, and then work the foal alongside the mare.

Stroke the foal with a brush a few times. Pick up one or two feet and pat the hoofs. Take a rope or a saddle blanket and drag it over the back. Standing alongside the foal, reach back and scratch under the tail. Put a finger up one nostril and wiggle it. Rub one ear. Lead him around a little more. Turn him loose.

Next time, do some other things. If you find an area inadequately desensitized, repeat the original habituation procedure until it is desensitized. This shouldn't be necessary if you did things right the first time. At another session, maybe load into the trailer, or hose the foal's legs down with a water hose, or handle some areas skipped last time.

Testing her mouth, nose, and face; all are desensitized.

Sensitizing Procedures

Interspersed with the desensitizing reinforcement procedures, we should occasionally do sensitizing reinforcement sessions. Again, all we'll do is repeat the original procedure, perhaps repeating it again in order to obtain enhanced reaction. When that occurs, stop. Stopping is a reward, and it helps to fix and enhance the desired response.

Remember that what we accomplished originally was to sensitize the foal to certain stimuli. By encouraging rather than suppressing evasive responses, we now have the foal sensitized to halter pressure, so that he will tie and lead. He is conditioned to follow us; therefore we don't actually need to create halter pressure to get him to follow us. He does so before he feels the pressure. We use the pressure only as a back-up if we don't get a response to our signal.

Similarly, we can control the hindquarters. Merely touching the flank will move the hind end laterally in the opposite direction. We have conditioned the foal to back up or to move forward on command.

Now we are ready to reinforce these responses. We needn't do all of them, and we can mix them in with the previous reinforcing procedures. For example, we said that we quietly haltered the foal in the pasture and stroked him. By doing so, we reinforced the *desensitizing* procedures.

Then we lead the foal to the right, to the left, and to the right again. We back up a few steps. Now we are reinforcing the *sensitizing* procedures. Move the hind end a bit to the left. Good! Now a bit more.

Each time we reinforce and enhance the response, which has become automatic.

Movement is the horse's way of escaping from danger. When we sensitize, we encourage movement in the direction we desire by allowing it, until the movement becomes a conditioned response to a specific signal or stimulus. Horses can be trained to race, cut, jump, or maneuver simply by directing flight.

Imprint-trained foals make good patients.

As training proceeds, the bond between the horse and trainer will strengthen. Here, Bill Riggins rides this young horse with only a halter, while interspersing trail riding with arena work.

15 PREVENTING PROBLEMS

WHEN I first started advocating the training of baby foals around 1970, there were very few people who responded to the idea. I would demonstrate the technique to my clients using one of their newborn foals. I didn't charge for this service be cause, admittedly, I had a very selfish motive. Imprint-trained foals make good patients. They are much easier for the veterinarian to work with and handle. The foals that I used for demonstration purposes, as well as my own foals, always turned out to be good patients.

Despite this, a majority of my clients would not incorporate the technique into their training programs. They seemed in-

Inprint-trained foals make good patients.

Inquisitive foals nibbling on yucca blossoms.

terested and even impressed by what I did, but the next year, their foals were as wild and flighty as ever. When I asked why they weren't using the training techniques I had shown them, they'd say, "Oh, we just don't have the time;" or "We've just never done it that way before;" or "I can't remember exactly how you did it"

In 1985, I made a video film called *Imprint Training of the Foal*. I showed the film to our clientele and started to use it at training seminars; from that point on, the technique caught on and has attracted worldwide interest. Within my own practice, large numbers of foals became easy to handle as the farms and ranches started to imprint train every newborn. I was able to identify these foals as soon as I started working with them, usually when they were wormed and vaccinated at 2 to 3 months of age. Their behavior and attitude were distinctly different from the foals who were conventionally handled. In the ensuing years, reports have come to me from all over the world, consistently de scribing successful results with all kinds of horses. Constant requests for a book on the subject led to the one you are reading.

I have, however, received a couple of reports of bad results, and I have had a number of people who expressed a fear of bad results. I want to address those concerns now.

I showed the clients of my former practice (The Conejo Valley Veterinary Clinic of Thousand Oaks, California.) the video film at a seminar for horse owners. As I said, many viewers then adopted the technique which I had been promoting with little success for 15 years. I don't know why the video was the turning point. Possibly the techniques are more clearly visible in a video than they are in a live demonstration.

Maybe the repetition of technique seen in the video is a superior teaching tool.

Repetition is one of the keys to learning. I have used the same teaching principle in this book. Or maybe in our culture, television has shaped our ability to learn so that it has become a more effective format. In any case, the idea caught on, and the next spring it was widely used in our practice.

Then, two clients telephoned us. One was a professional trainer with a month old Quarter Horse colt. The other was an amateur owner with a month-old Thoroughbred colt. Each of these ladies had, they said, done the foal training procedure exactly as I had shown them in my video. But in both cases, the foals had turned out to be outlaws that could not be handled at all or touched anywhere on their bodies.

What had gone wrong? It took me a while to figure it out. These were both super-dominant colts. The trainers had been in a hurry. In my film, I repeatedly said that most foals will habituate to about 30 sensory stimuli, but to persist in the stimulus until after habituation occurred, regardless of how many repetitious stimuli were needed.

What the two owners had done was stop each stimulus, even though the foal was still struggling to escape it, at the count of 30. Instead of desensitizing the foals, they sensitized them, and the foals had become untouchable.

We tied the foals' legs while they lay on the ground, and I had the owners re-do the imprint-training ritual, repeating every stimulus 100 times. When the foals were allowed up, they were gentle and submissive.

Therefore, when doing this kind of training (habituation) in a horse of any age, persist in the stimulus until all fear is gone, or you may cause more harm than good.

Unjustifiable fear of bad results has deterred other horse owners from at tempting imprint training. These concerns have been expressed by experienced horsemen, and invariably they have been older men, which may or may not be significant.

These people are afraid that a foal trained as I advocated will grow up to be a spoiled pet with no respect for humans, that it will bite people and, very commonly, that it will be a lazy clod unsuitable for racing or performance events such as reming or roping.

It is true that baby foals who are coddled and petted and handfed often turn out to be spoiled pets, but it isn't because of early training. It is because of *improper* early training and handling. As one example: A foal should never be allowed to suck or nibble on your fingers or shirt buttons as this can lead to biting.

As the foal starts to grow, I teach him to respect my personal space, to defer or yield to me, to give way, and to stay back a bit. Otherwise, when frightened, for example, the foal might try to climb into my lap, or on top of me. This is what old timers fear in the "spoiled pet."

The training I have described in this book makes gentle foals, but they are not spoiled foals. They are responsive. They are submis-

Checking out a feed tub to see if it's edible.

sive. They are obedient. A gentle, responsive, submissive, obedient, and well-mannered foal is not spoiled.

I try to teach foals not to fear harmless things, no matter how frightening they might seem. I teach them to lightly respond to certain signals: to lead, tie, back, and rotate the forehand and hind quarters. I try to teach them good manners. I teach them:

"I can touch you, but you must never touch me."

"I can put my hand in your mouth, but you do not ever put your mouth on me."

"I can touch your feet, but your feet must never touch me."

"I can rub your face, but you must never rub your face against me."

"I can get on your back, climb all over you, and push you in any direction, but you must never do those things to *me*."

There is a big difference between being gentle and being spoiled. The problem is that many people see two potential extremes in the attitude of a horse toward a human. At one extreme they see contempt; at the other, fear. There is an in-between relationship that is one to strive for, called respect. Respect with trust. It can be obtained in a newborn foal with the behavior-shaping procedures I have described.

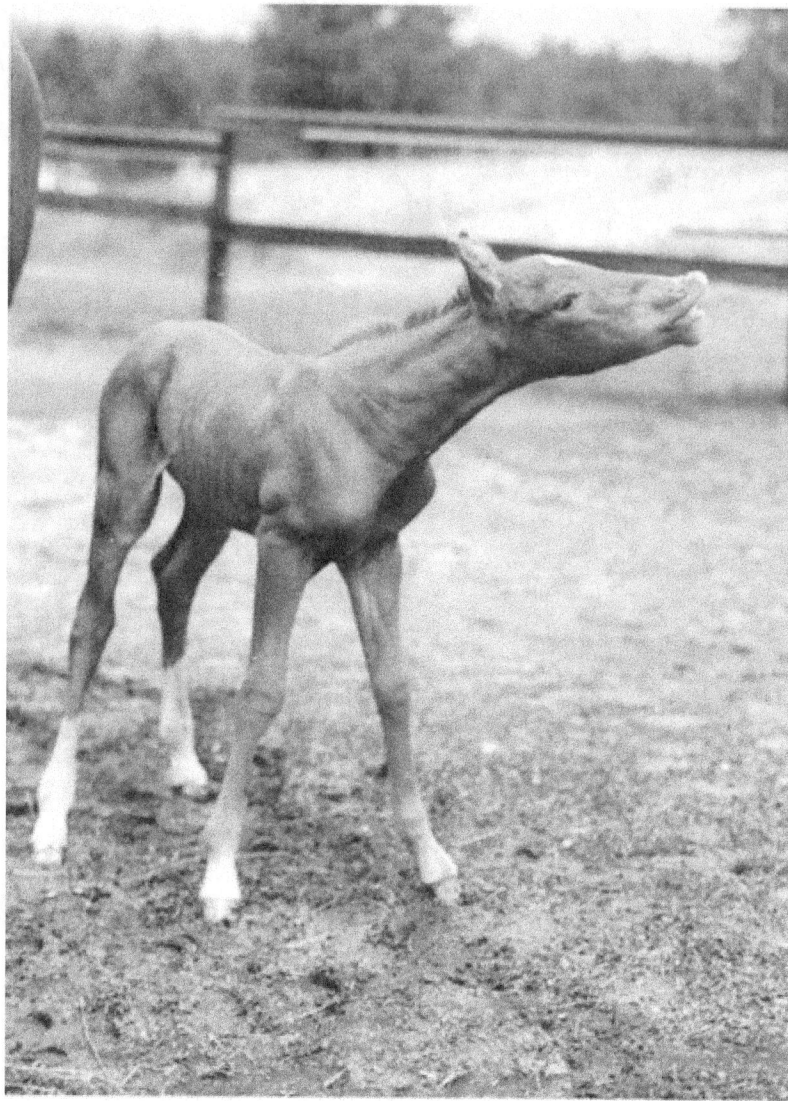

However, there are occasional times when a foal might need disciplining to discourage undesirable behavior, such as nipping or biting. But keep in mind that foals are inquisitive, and nibbling or chewing is their way of checking some thing to see what it is—just as a human baby puts things in his mouth. You don't want to discipline a foal for this, so the best solution is not to put yourself in a position where he can nibble on your buttons, shirt sleeves, or whatever.

If a foal flat-out bites, an immediate slap to the nose will usually discourage him from trying it again. But the slap must be immediate; not 30 or 60 seconds later. And just one slap is all that's necessary. If the slap is not delivered immediately, it simply teaches the foal to bite and then duck out of the way.

In order to extinguish any undesirable behavior, the punishment should, ideally, be simultaneous with the action; or, at least, follow so swiftly to be nearly simultaneous.

Extinguishing undesirable behavior, especially if it is habitual, with an aversive (painful and/or frightening) stimulus is called counterconditioning. It is extremely effective in horses. (My video, *Influencing*

The training I have described in this book makes gentle foals.

the Horse's Mind, demonstrates several examples of counterconditioning techniques. Visit www.robertmmiller.com to order.

As said earlier, a single slap is all that's necessary. Repeated punishment will only serve to make the foal head-shy, and after the first slap, the foal will forget why he is being punished.

In very young foals, the best way to discourage nibbling and biting is to flick the foal's nose with a finger the instant the muzzle approaches you. Remember, we want to teach the foal we can touch him, but that he cannot touch us. We can touch his lips, or mouth, or nose, but he cannot touch us with them. It is bad manners.

What I do is look away from the foal and flick his muzzle sharply, with no arm or hand motion. I just flick my finger, with a snap, at the muzzle. I don't want the foal to fear me. By looking away when I flick the unsuspecting foal, he doesn't fear me. He has punished himself for doing some thing I don't want him to do.

Also keep in mind that foals, out of sheer exuberance, often kick. They usually do not mean intentional harm; they are simply having fun. The answer: Stay well away from a loose foal while he is playing, or when you think he's apt to wheel and kick.

If you have a foal on a lead rope, stay away from his rear end so he can't kick at you. But if he does manage to kick, or strike at you with a front foot, whack him once, immediately, with the lead rope. Again, that's usually all that's necessary.

For a foal with a definite inclination to kick, a crop or whip is more effective than a halter rope. A whip serves as an extension of the hand, and should be about 4 feet long. A dressage whip is ideal. Just as with nipping, the kicking should be punished by an immediate, single, sharp sting of the whip.

Suppose the foal is loose in a stall or pen, you are trying to catch him, and he keeps turning his tail toward you. You are right to assume that he might kick at you if you get too close.

Yet foals, as well as horses of any age, can be taught to respectfully face you. You do this by rewarding the foal, when he faces you, by assuming a passive, non threatening stance, and by stepping back, away from the foal, while still facing him. When the foal turns his rear end toward you, threaten him by assuming an assertive, alert posture, and moving toward him. Heighten your threat by waving a whip or halter rope.

This must be done in a confined space. The *instant* the foal shows the *slightest* indication of facing you, reward him by assuming a passive stance and stepping back. If done properly, this technique always works.

With variations, this is the same method that Pat Parelli, Ray Hunt, and many other capable horsemen use to quickly teach a horse to face them, and to approach them.

I want to stress that what I have just discussed is not a regular part of my baby foal training program. Imprint-trained foals will usually walk up to people, and are respectful. However, a foal that has not been imprint-trained, and that is real spooky or dominant, might need this kind of training.

Imprint-trained foals will usually walk up to people and are respectful.

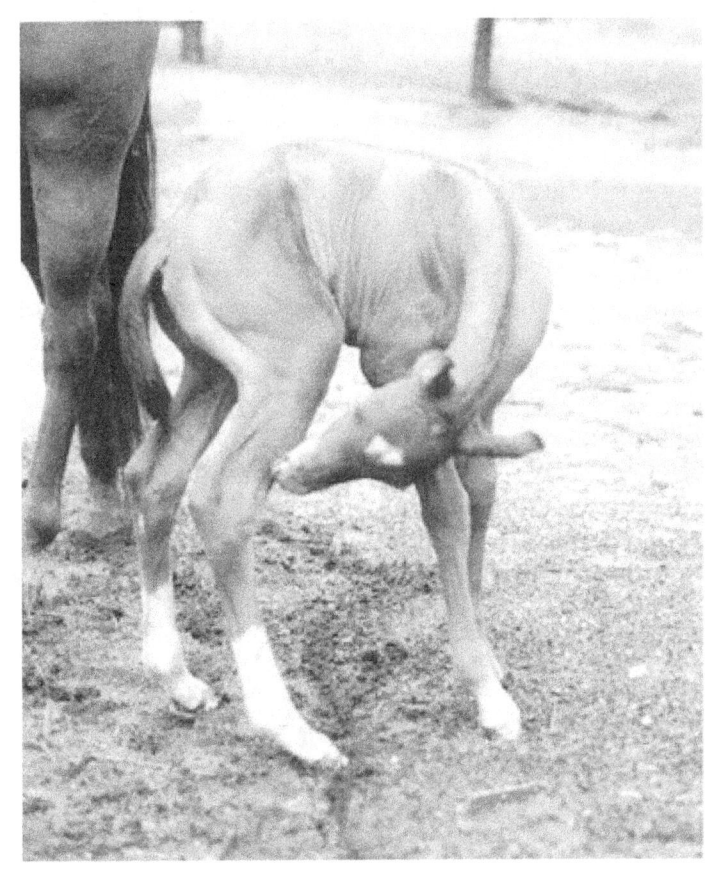

16 THE RICE HORSE

OF ALL the horsemen resistant to the idea of training foals to be gentle and well mannered, the most hesitant have been the race horse people. There is a prevalent belief in North America that if a young horse is taught to be well-mannered and obedient, and that if he is not flighty, he will not run well. This misconception is the reason most of our young American race horses are hard to handle. It is the reason so many of them injure and kill themselves when being shod or doctored, when being trailered or vanned, and pulling back if tied. Thousands of race horses have suffered neck injuries, ending up as "wobblers" because they were not properly taught to tie and lead. In other words, they were not adequately halter-broke.

Two experiences in my life have taught me the fallacy of not properly training race horses.

In the 1940s, I worked at the starting gate at Rillito Racetrack in Tucson, Arizona. I saw men and horses injured because the horses were so excited and flighty at the starting gate. I saw bad starts which lost races (a bad start is especially harmful in a sprint, and Rillito was primarily a Quarter Horse track), and which caused accidents.

In my youth, I used to rope calves. A really good calf roping horse (I never owned one, but I saw a lot of them) stands quietly in the box, like a coiled spring, waiting for the calf to be released. When he comes out of the box, he does so faster than any Thoroughbred comes out of the starting gate. In those days, most calf ropings used small Brahma calves that could run at very

Two experiences in my life have taught me the fallacy of not properly training race horses.

Malibu Valley Farms, a Thoroughbred breeding farm located in Calabasas, Calif., has been imprint-training every foal born on the ranch since 1983. Many of these foals—such as African Storm, shown here winning a 6-furlong race at Santa Anita in 1989—have gone on to become top race horses.

Kansas City, another Malibu Valley Farm imprint-trained foal who became successful on the track. This winner's circle photo was taken at Hollywood Park in 1989.

high speed. It takes a tremendous start and immediate acceleration to top speed to catch up with such a calf.

Good Quarter Horses, burdened with a heavy stock saddle and rider, can do that in *2* or 3 seconds, even when the calf is allowed a generous head start. Significantly, they do it from a standing start, and the best ones do it without showing any crazy behavior.

There is no reason that a race horse cannot be trained to enter a starting gate quietly and sanely, then gather himself for the race, and come out like a rocket being launched when the gate opens. Rodeo horses do it all the time.

Race horses don't run because of fear. They run because they want to run. Some of them are physiologically and anatomically built to run fast. If they can run fast and they want to outrun the other horses, you have a potential winning race horse. Flightiness and bad manners have nothing to do with it. Indeed, they limit the racing potential for many horses. The fearful horse wants to stay in the center of the pack where, in the wild, he would be the safest. Moreover, the frightened, excited horse often has spent his energy and "adrenalin rush" before entering the gate.

Anyone who has seen young horses in pasture has observed that they frequently race enthusiastically. Each species, when it is young, engages in instinctive play which is based upon inherited behavorial characteristics, which in the wild will help ensure survival after maturity. Thus, puppies chew, bite, and tug. Kittens stalk, charge, pounce, and maul their playthings. Human toddlers hit and throw things—skills which were necessary for the survival of early man.

Horses, a grasslands prey species, rely upon a burst of speed to escape danger. Foals, therefore, love to sprint, and they do it in groups because, in the wild, they are a herd species. The sprinting is often interspersed with bucking, kicking, and striking, all actions which can deter the predator who manages to outrun a horse.

Horses raised in an environment which allows long sprints together with peers are psychologically conditioned to race. However, even in a group of young foals, there is a dominance hierarchy, and some foals learn to *not* run in front where they would challenge the natural leaders. They learn, instead, to run with the pack, back where it is safe.

Of course, some foals run behind simply because they are slower than others, lacking the speed and stamina to lead. But for the purpose of this discussion, let's assume that all of the foals have equal speed, and that the only variant is their *desire* to be in the lead. Supposing that, it seems to me that in order to get a naturally submissive horse (a follower rather than a leader) to move to the front, he must be *motivated*. The jockey's whip, obviously, is one method of providing such motivation.

Now, granting that there is skill in volved in the use of the whip—when and how to use it, to enhance the horse's stride rather than interfere with it—every jockey knows how to use the whip. Why then, do certain jockeys have the inexplicable ability to get horses to win races? I believe that their ability is akin to that of other great riders, be they dressage masters, steeplechasers, or cutting horse riders. In addition to their obvious skills, they communicate with the horse. Such communication must involve an attitudinal factor on the part of the horse; a willingness to try for this person, a willingness to be led.

It follows, therefore, that the people oriented horse will race more effectively than the horse that is skittish and afraid of humans in general. When I worked in the starting gate as a youth, I noted that the frightened horses (which acted badly and dangerously in the gate) not only tended to start poorly, but also liked to run in the center of the pack where, I suppose, they felt safer.

So, I believe that imprint training will *enhance* rather than handicap a foal's racing future.

When I was making my first foal imprinting video, I used Thoroughbred foals for most of my subjects, on the farms which were responsive to the concept of early training. One such farm had two co-owners. One was an American, whom I had met. The other was an Irishman, whom I had never met. As we filmed a newborn foal and trained it, the Irish partner arrived coincidentally on a visit from overseas.

"What's happening here?" he wanted to know. I introduced myself and, apprehensive that he would disapprove of my "spoiling" a very valuable racing foal, hastily explained the concept of early training. I told him that the farm, in fact, had been using the technique on all of their foals for 2 years and hoped that he would not mind what we were doing.

"You see," I said, "Most race horse people think that if a horse isn't flighty that he won't run, and that's a fallacy."

He was silent a moment, and I was concerned about his feelings. Then he said, "Let me correct you, Doctor. Most *American* breeders subscribe to that fallacy. It is not a worldwide opinion. In Ireland, I assure you, we work with our foals from birth on and teach them manners. I have seen the Thoroughbred industry all over the world, and the idea that horses must be ill-mannered in order to run well is prevalent in only three countries that I have been in: the United States, Canada, and Australia. I applaud your efforts."

Greatly relieved, I asked him why he supposed that those particular countries seem to oppose the proper training of young foals.

"I don't know," he responded. "Perhaps it is an expression of the frontier mentality."

I once visited the Thoroughbred training center in Chantilly, France, and was impressed with the discipline and good manners shown by the young horses on the training track.

I find it ironic that the most monetarily valuable horses of all, our racing Thoroughbreds—possessed of athletic power and speed which increases the chance of injury to themselves or to the people work-

In any group of horses, many prefer to run in the middle of the herd, where it is safest. These are Arabians on Rush Creek Land and Live Stock Company, located at Lisco, Nebraska. This ranch is well-known for the endurance horses it raises.

ing around them—are the least likely to receive early training. It should be the other way around. *Because* they are so valuable; *because* they are natural athletes possessing fast responses, speed, and power; *because* they are exposed at a young age to the excitement of the training track, the van, the sales ring, and the racetrack; *because* they must frequently be seen by the farrier, the veterinarian, the trainer, and the groom; *because* of these reasons, Thoroughbred foals should be trained from the moment of their birth to be well-mannered. They should also be trained not to fear harmless, but frightening stimuli. They should learn to respect man, to be responsive to his signals, and, yes, to bond with him.

Thoroughbred breeding farms that have adopted imprint-training methods produce winning horses, and the injury rate to both their horses and humans working with them has been greatly reduced. One ranch, Malibu Valley Farms of Calabasas, California, has imprint-trained every foal born on the ranch for 8 years at the time of this writing. They have had their share of winners, racing in both North America and in Europe.

The same was true of Oakbrook Farm of Canoga Park, California. They imprint trained every foal born on the farm for many years, producing successful race horses until the farm was sold.

17 MULES

MULE FOALS respond to imprint training as well as horse foals do. The response, in fact, is probably more dramatic because mules in general are more difficult to train than horses, being keenly perceptive, less predictable, extremely self-protective, and far less forgiving.

The mule is a hybrid creature, the result of breeding a jackass (a male donkey) to a mare. Mules have 63 chromosomes—31 from their sire and 32 from their dam. The only obvious result of the inequity is sterility. With rare exceptions, mules do not reproduce, even though they cycle normally and will breed. Male mule foals should be gelded as early as possible. I recommend that castration be performed before they are a month of age, and have done it as

The procedure for early training of mule foals is similar to that for horse foals.

Although I always teach horse foals to lead first, before tying them, I usually teach mule foals to tie first. Horse foals will often throw themselves if tied before being taught to lead, but mule foals are too clever to do that, as a general rule. This mule filly is 15 hours old. I have passed the lead rope around the inner tube, and stand behind her so she cannot pull back too hard. After two tries at pulling back, she learned to jump forward to escape the halter pressure.

When she learned to jump forward to escape the pressure, I stepped aside. She pulled back one more time, and then jumped forward.

That was it! She did not pull back again. I tied her to the inner tube with a slip knot and watched her for 10 minutes.

young as 1 day of age. Female mules may be similarly neutered, although most female ("molly") mules *are* less temperamental than mares during the estrus period.

The procedure for early training of mule foals is similar to that which I have described for horse foals. Because of their intelligence and strength, mule foals can

IMPRINT TRAINING

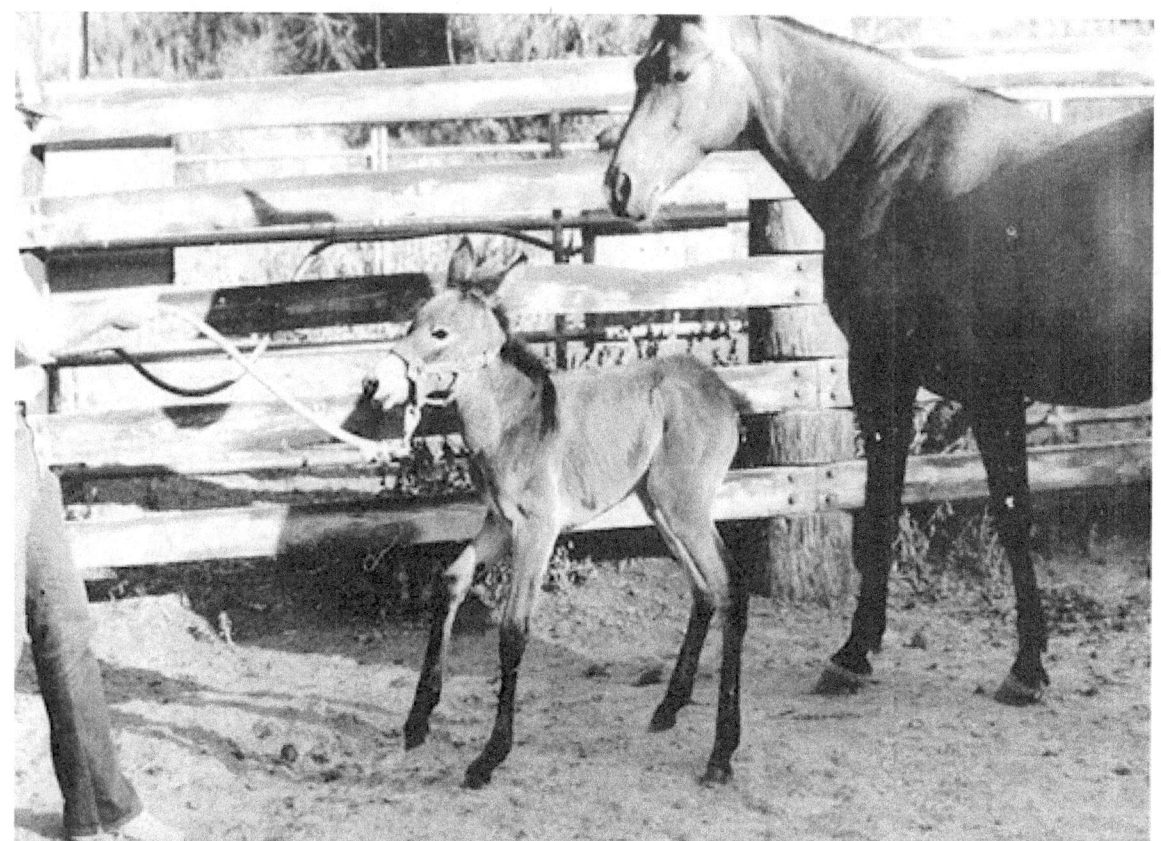

When she learned to jump forward to escape the pressure, I stepped aside. She pulled back one more time, and then jumped forward.

That was it! She did not pull back again. I tied her to the inner tube with a slip knot and watched her for 10 minutes.

...she bounces forward. I paused for a moment to let her think about it, then repeated the process.

She's leading-not smoothly, but with a series of little jumps forward. In less than a minute this smoothed out so that she was leading from a walk. Then I turned her loose-and repeated the entire lesson after 3 days. At that time, I used a butt rope to encourage her forward when she hesitated. At 12 days, I was ponying her from her dam, and she trotted alongside.

be imprint-trained more swiftly and more easily than most horse foals—at least in my experience. This is fortuitous because an ill-mannered or unbroke mule is a handful, indeed.

I have imprint-trained only one hinny. A hinny is the result of crossing a stallion with a jennet—a female donkey. It responded exactly as mule foals and horse foals do. Hinnies are rarer than mules simply because few stallions will breed jennets, whereas many jackasses are willing to breed mares.

In order to understand mules, one must realize that they are not horses. They are the hybrid offspring of two different species: the horse and the ass. These species are related, but they are not the same. The horse is, in the wild, an inhabitant of grassy plains. Its primary survival behavior is high-speed flight. Most varieties of wild ass, by contrast, evolved in arid, rocky, mountainous terrain. An exception is the Somali wild ass, an endangered species which is normally found in flat desert wastelands in sub Saharan Africa, and which is capable of very high speed.

Like other prey creatures which live in precipitous terrain, the wild ass, when frightened, chooses between a choice of reactions. He can remain motionless, realizing that he is not vulnerable in an inaccessible location. This tendency to "freeze" is the source of the donkey's and the mule's reputation for stubbornness.

Alternately, the ass can decide to flee, just as the horse does, and this decision can be made almost instantly. However, even though mules and donkeys can often resort to flight at horse-like speeds, they usually do not run as blindly as a horse, and they will "think" along the way. This is the reason that mules and donkeys will rarely run through a wire fence or off a cliff as a horse will.

Lastly, the ass may choose to attack, something the wild horse will rarely do unless cornered. Mules and donkeys, therefore, will not hesitate to attack dogs, and they are capable of killing even large dogs. In the early colonial days of California, jennets were run together with the broodmares on open range in order to protect the foals against wolves. Burros have also been used to protect sheep against coyote predation.

Mule foals must be carefully accustomed to dogs. Even mule foals a few days of age might attack dogs. However, they can be desensitized to dogs, and even learn to bond with them by keeping them next to dogs, separated by a mesh fence. Weanling mules, if pastured with goats, will learn to tolerate dogs. I questioned the validity of this bit of folklore, until I tried it and found that it worked.

Many of the problems which traditionally are expected in mules may be completely eliminated by imprint training as soon as they are foaled. The problems I refer to are:

1. Mules resent having their ears touched.

During the initial imprint-training session, spend extra time doing the ears. Be sure to include the ears during later reinforcement sessions. Imprint-trained mules do not object to having their ears handled.

2. Mules are hard to shoe.

Imprint training, in my experience, completely prevents foot-handling problems. The mules I have raised all readily pick their feet up, and allow trimming and shoeing.

3. Mules tend to pull back and run off.

After being trained to lead and tie at 1 day of age, my mules have not developed these vices. However, young mules should *always* be led with a long lead rope and some sort of halter which will punish them the first time they try to run off. They *will* try, sooner or later, and it is even more important to not allow this to happen with young mules than it is with young horses.

Although mules and donkeys often flee at horse-like speeds, they do not run as blindly as a horse.

Floating the teeth on Jassper, my regular riding "horse."

Similarly, when tying young mules *or* young horses, it is a good idea, whenever possible, to tie them short and high to a rubber inner tube. I do this regularly until my horses and mules are 2 years of age, and continue to do so in older animals. There is no better way of teaching them not to pull back. Pulling back when tied to a stout elastic inner tube is an unrewarding experience.

Mules are superbly sure-footed trail mounts. In this regard, I believe they are superior to horses. However, because they are wary and so self-protective, they can be balky on the trail. For this reason, I think that it is especially beneficial to lead a mule foal alongside the mare as de scribed in a previous chapter. If, early in life, the foal learns to cross streams and bridges, go up and down steep slopes, tolerate traffic, noise, and livestock, and accept varied surroundings, he will be less inclined to balk under saddle later on.

Mules, generally, are even more sociable than horses, and, as a result, are more inclined to become herd-bound or barn sour. They might refuse to leave the stable area or, once away from home, they might want to return. If the rider has a good relationship with the green mule (bonding plus submissiveness), the mule will be more tolerant of the strange surroundings.

It also helps if the ride takes varying routes. Sometimes, when returning to the stable, it is a good idea to keep right on going past it and ride off in another direction.

Lastly, when finally the ride is over, don't do pleasant things such as unsaddling, feeding, or bathing. Instead, make the homecoming a bit unpleasant by working the young mule in the arena, or bitting him up for an hour, or simply tying him up for an hour or two.

The tendency towards being herd bound is minimized by leading the very young mule foal, from the mare, some distance from home and leaving them both there for a period of time. The foal learns that being away from home is not a catastrophe. Later, after the foal is weaned, he can be led away from home on a reliable, well-mannered lead horse (or mule).

If possible, do this on different lead animals each time. These techniques are beneficial for horse or mule foals, but since mules are more inclined to be herd bound, it is especially valuable for them.

18 EFFECTS OF IMPRINT TRANING ON MARES

ONE OF the most frequently expressed concerns regarding early foal training is that it will somehow interfere with the bonding between the mare and the foal or that it will cause the mare to reject the foal.

I have been experimenting with what I call imprint training since 1967. I have done all of my own horse and mule foals since that year plus many, many dozens of foals for my clients. I would estimate that I have personally done a couple of hundred foals, of all breeds. In addition, my clients have done many hundreds more by themselves.

When I released videos on these techniques (Early Learning - www.robertmmiller.com), they attracted worldwide attention. In 1987 I retired from a 32-year veterinary career and since then have been largely occupied teaching equine behavior and, especially, early foal training. At

I have never seen a mare with an imprint-trained foal act aggressively toward humans.

Range mares grow up seeing lots of new foals.

my seminars, I regularly poll the audience to find out what kind of horses they are involved with, and whether any of them are already utilizing imprint training. At the time of this writing, from 10 to 12 percent of my audiences are already using the method on their foals, and about half of them have used it for more than one foal crop. They are universally satisfied with the results.

I always ask if anyone has ever experienced a case of a mare rejecting an imprint-trained foal (I never have). Thus far, one case was reported where the mare wanted the foal, but would not allow it to nurse, no doubt due to udder pain. She eventually accepted the foal. Of the thousands of foals accounted for by these audiences, there has not yet been one case of foal rejection. I am slowly coming to a conclusion that only time will verify:

Foal rejection can be prevented, in most cases, by imprint training.

Why? A *majority* of foal-rejection cases occur because the mare is afraid of the foal. The most common offender is a first foal Arabian mare. Arabians are a very sensitive breed. Because of their value, most Arabian mares don't foal out in pastures. Range mares grow up seeing lots of new foals. Today, a lot of "hot-house" horses never see a newborn foal until, suddenly, one appears in their own stall.

Such a mare acts alarmed when she sees the foal. She acts the same way horses do when they see a pig for the very first time, or a burro, or a peacock. They're scared.

Now, some mares refuse to let foals nurse because of udder pain, and they may even kick and injure the foal. We're not talking about that sort of rejection. Nor are we talking about the occasional freak reaction where a mother (of any species) will attack and kill its newborn young. We're talking about the more common panic reaction that some mares show when they get to their feet after foaling and suddenly see this weird little creature in their stall or paddock. Their desire, then, is only to *avoid* the foal.

I believe that, when such a mare looks around, and she sees a familiar human restraining and manipulating the foal, it *reassures* her. Typically, she will soon tentatively come up and sniff the foal. Then she'll taste (lick) it. That's all it usually takes. She's

The majority of foal rejection cases in my practice occurred in first-foal Arabian mares that had never seen baby foals.

hooked. Soon, the foal is bonding with and imprinted with both the human and the mare.

I believe that's the reason I have not yet received an authenticated report of a mare rejecting an imprint-trained foal.

Inevitably, it will happen, but when one considers the statistics, I think that it is safe to say that most cases of foal rejection are preventable, especially when one considers that from 1970 until around 1985, our broodmare practice was predominantly Arabian.

I have noticed another thing about imprint-trained foals. The mares, the next day, don't show aggressiveness toward humans. It is very common for even usually gentle mares to show extreme protectiveness toward their foals. Many of these mares will viciously charge anybody who gets near the foal. Many times I have had to rope a mare over a corral fence before I risked stepping in with her to see her new foal.

I have *never* seen a mare with an imprint-trained foal act this way. At my seminars I always poll the audience to see if anybody else has seen one. Thus far, there hasn't been a single one. Why?

I suspect that, when the mare gets up after the foal is born, that the smell of the placental fluids is on my hands, as well as on the foal. This smell triggers a maternal instinct in the mare. I believe that a three-way bonding results: Mare to foal to human.

Obviously, this whole subject deserves further investigation and needs further understanding. The important thing, that I can say now, is that rejection of imprint-trained foals by their dams is not a problem.

Of course, I'm talking about well mannered, halter-broke mares. Mares that have had little human contact and are not, at least, tractable and well halter-broke are a different story. Their fear of humans will be transmitted to the foal. As I've said many times, no mare should be bred until she is halter-broke and well-mannered—and, ideally, broke to ride. Good horsemen aren't proud of unbroke or ill-mannered broodmares.

I believe that when a mare sees a familiar person working with her newborn foal, it reassures her that "this weird little creature" is okay, and she will accept it.

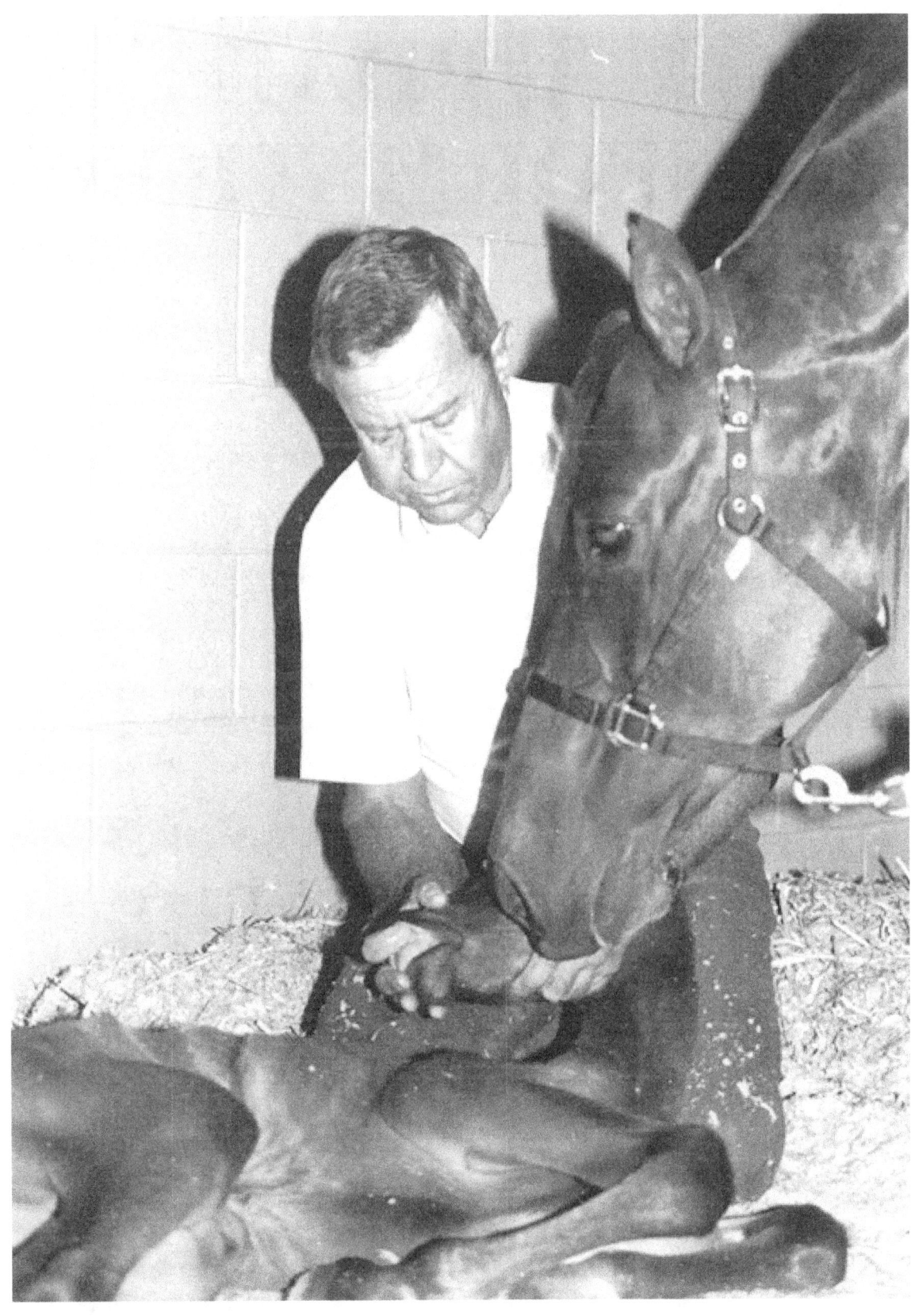

19 UPDATE INFORMATIOMN

Reprinted from *Western Horseman*, April 1994.

TRAINING THE newborn foal accomplishes four things: bonding, desensitization, sensitization, and dominance—all of which are explained in *Imprint Training of the Newborn Foal*. However, I have become aware of several mistakes some people are making in each of these four categories, as well as in the safe positioning of foal and handler. I'll briefly describe the areas and then detail what should be done to pre vent mistakes.

1. Bonding. If imprinting is done within the first hour following birth, the foal will develop a powerful bond for the person doing the training. This occurs because contact has been made within the imprinting period and the foal will want to follow and respect the human involved, just as he will his dam.

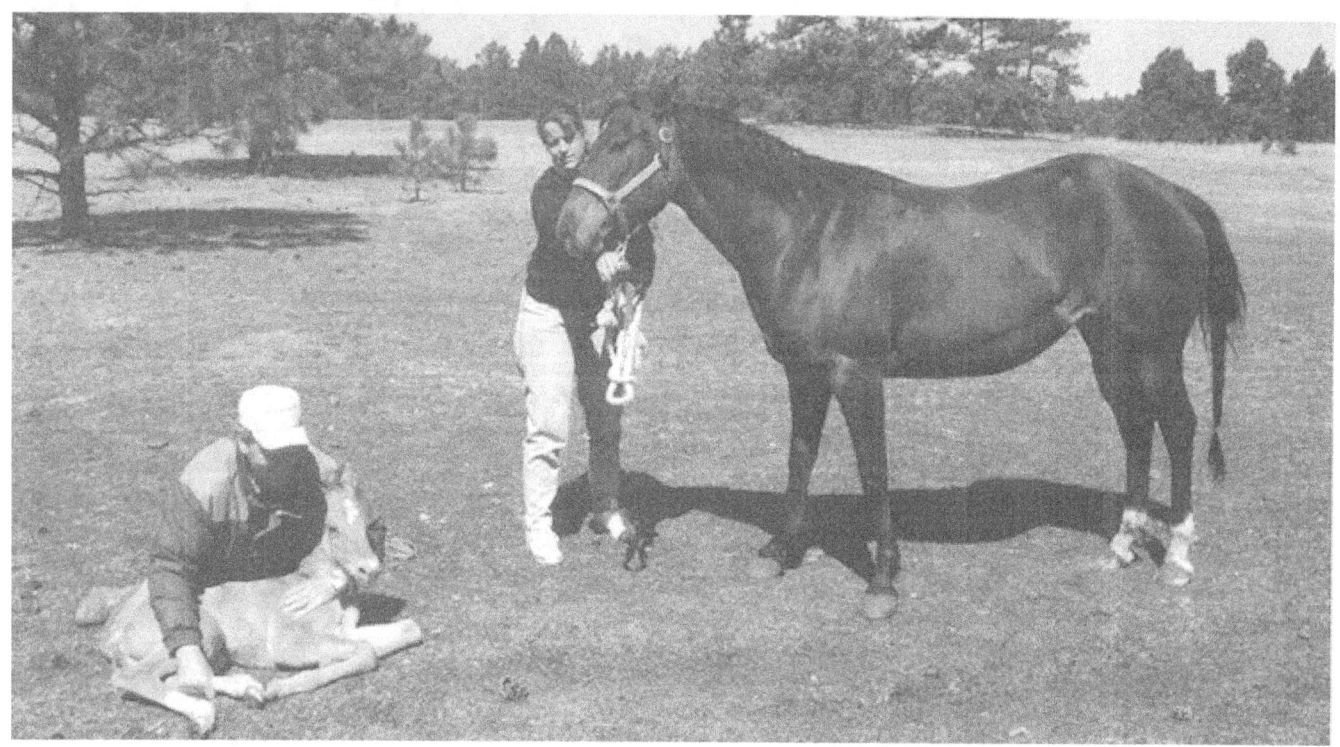

This is the ideal position for imprint training. The mare handler has the mare under control, and the foal and mare are facing each other. The foal's legs are folded so that the mare can't step on them. The foal handler is in a safe position behind the foal. The mare can be allowed to move closer to the foal, so she can touch him with her muzzle. Ideally, the mare handler should be wearing boots to minimize the chance of injury if her feet are stepped on by the mare.

Photo by Kathy Kadash

UPDATE INFORMATIOMN

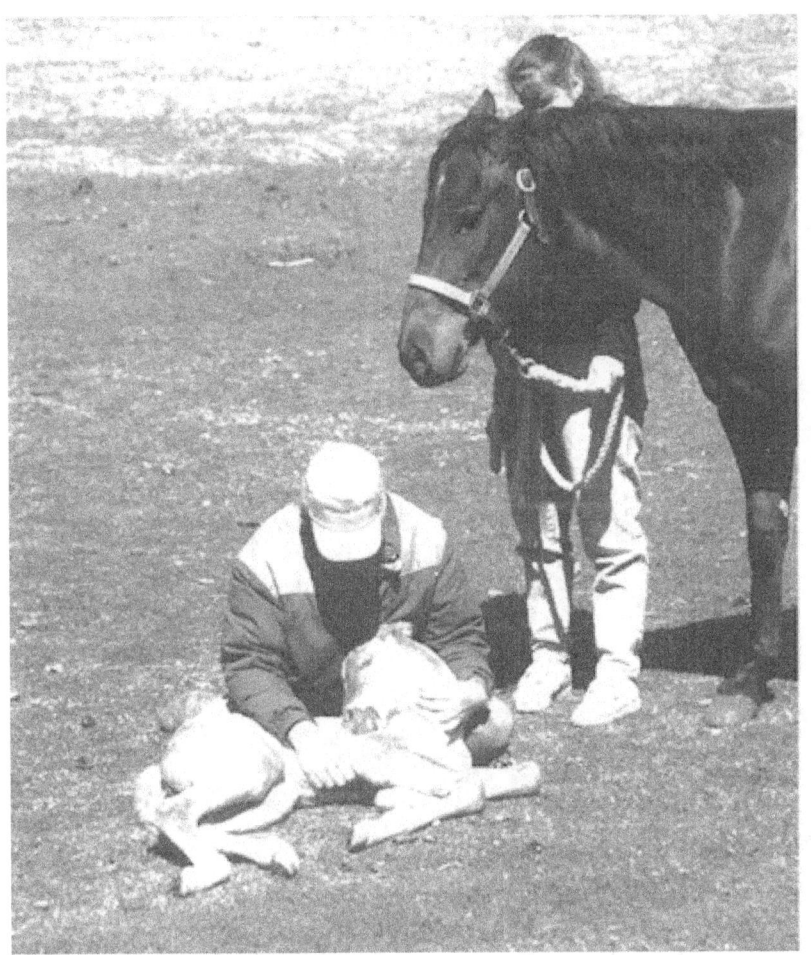

What's interesting here is to see the total lateral yielding of the foal's head. It will remain so until the handler straightens it. This paves the way for swift halter breaking the next day. Lateral flexions can be taught at birth.

Photo by Kathy Kadash

> *Following the publication of this book in 1991, many horsemen began imprint training their foals. In the vast majority of cases, the training has been accomplished very successfully. But, Dr. Miller has become aware of several mistakes people are making. So in our April 1994 issue of Western Horseman magazine, we printed an article by Dr. Miller titled "Update on Imprint Training." In the article, Dr. Miller pointed out several common mistakes being made, and how to avoid these problems.*
>
> *In addition, the March 1996 issue of WH contained another article by Dr. Miller titled "More on Imprint Training."*
>
> *We have now added both of these articles to the book as chapter 19.*
>
> —*Pat Close*
> *Editor Western Horseman Magazine*

Some people mistakenly think they are imprinting when they apply the imprint techniques on foals who are 1 or more days of age. You can certainly train a foal at any age, and the younger he is, the faster he will learn. You can also obtain bonding from horses of any age, but actual imprinting can only occur right after birth. People who have tried both methods can clearly see the difference in the foal's attitude and responses.

Some people also feel that if they are unable to imprint train a foal right after he is born, there is no point in applying the techniques later. This is not true. As I just mentioned, the younger a foal is when he is handled and trained, the easier it is. So do not hesitate to apply the imprint-training techniques in the first few days after a foal's birth; just do not call it imprint training.

2. Desensitization. A foal can be quickly and permanently desensitized to many things that would ordinarily frighten him later on.

It is imperative that the "flooding" stimuli be continued beyond the point of habituation (passive acceptance). You can't do too many stimuli, but you certainly can do too few. If you stop the stimulus while the foal is trying to evade it, you will *fix* that reaction. You will have sensitized the foal instead of desensitized him.

For example, if you are desensitizing an ear to being handled and you stop the desensitizing process while the foal is jerking his head away, he will have learned to jerk his head away whenever you handle that ear.

3. Sensitization. The foal can quickly and permanently be conditioned to respond to other stimuli. So, by the second day of life, the foal can back up, move forward, move laterally, lead, tie, and pick up his feet on command because he has learned to respond to pressure, not to quietly accept it without moving.

IMPRINT TRAINING

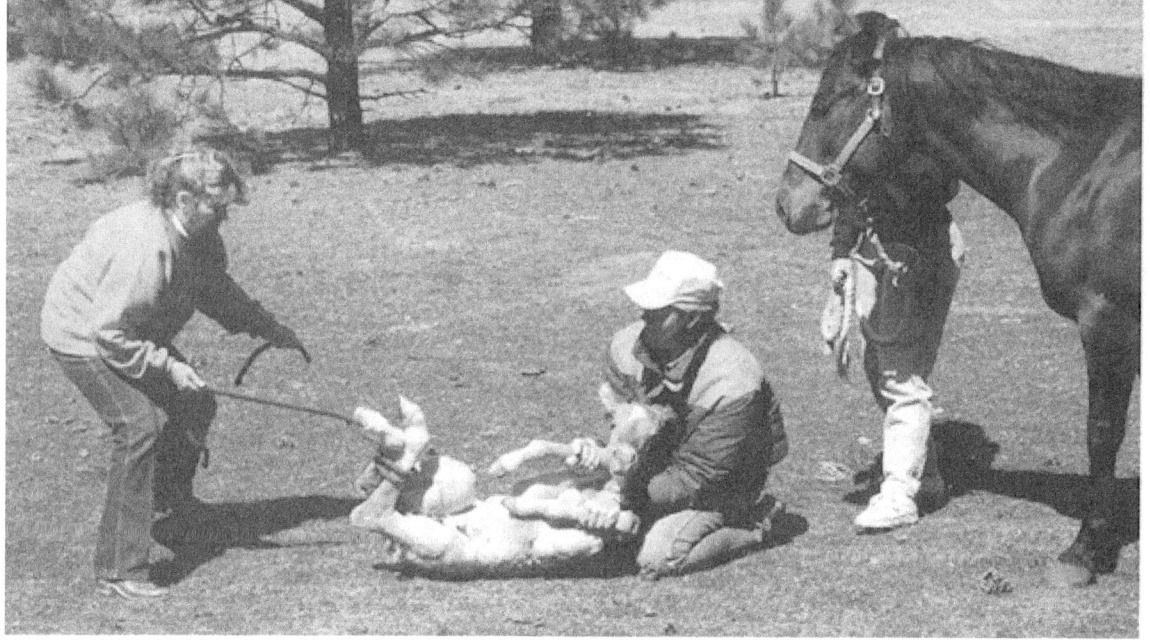

Here's a sequence of four photos showing how to turn a foal with the help of a third person. The foal handler should grasp both front legs by the cannon bones and position the foal's forequarters between his thighs as he kneels on the ground. The third person should loop a soft rope around the hind legs. Together, they should turn over the foal in one smooth move. Ideally, the foal handler should not have his back to the mare.

Photos by Kathy Kadash

The most common problem with sensitization is that some people simply don't try to achieve this, especially if they have heard about imprint training, but have never read this book or received formal instruction in how to do it. My training regimen does not simply consist of petting the newborn foal. I train him in many ways.

4. Dominance. Because a horse responds with submissiveness to anyone who controls his body position, the foal will quickly accept as a leader anybody who controls his ability to run away or, conversely, who can manipulate his body position.

Here is where most failures occur, and it's not simply because of a lack of understanding. Some people (especially women) cannot resist allowing the foal to suck their fingers or invade their body space—such as nosing or nibbling their clothing. The

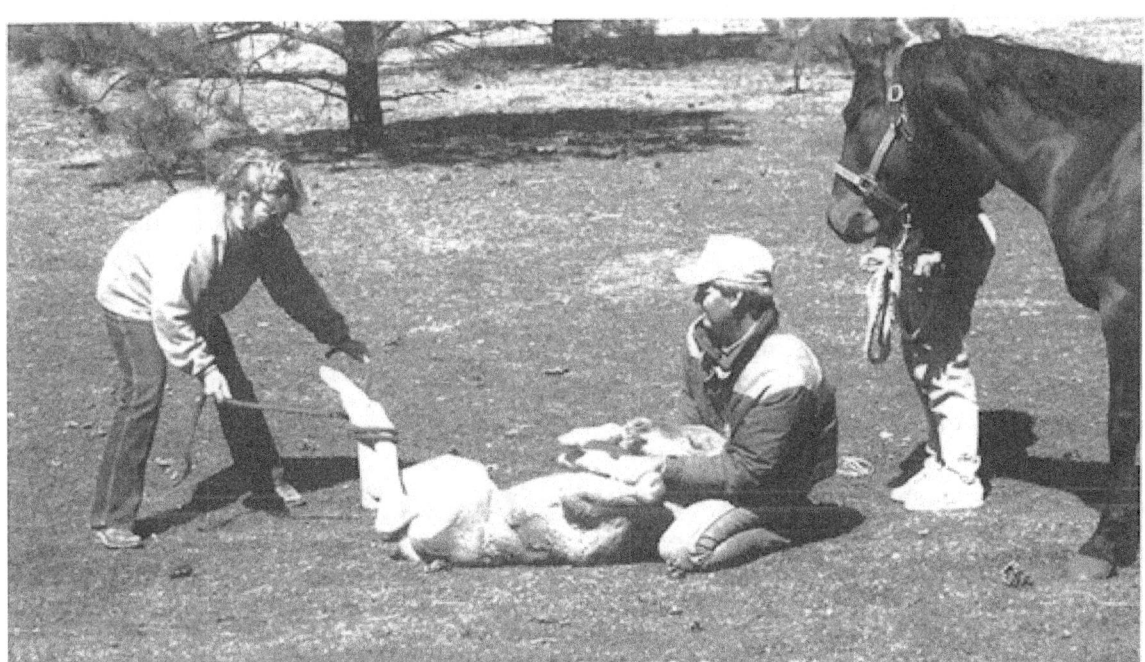

result is a spoiled, disrespectful foal—especially if the foal has a dominant personality.

With the first qualities established, we must also teach the foal the following:

A. I can touch you, but you may not touch me without permission. I can touch your mouth, but you must never put your mouth on me. I can pick up your feet, but you must never touch me with your feet. I can invade your personal space and be close to you, but you must never do that to me unless you're invited.

B. I can, from the moment of birth, control your body positioning. You can not escape if I tie you, or if I lead you.

You must back up, move forward, or move laterally any time I ask you to. I can immobilize your limbs and you will passively accept it. I am your leader and you will want to follow me and respect me. I will not hurt you and you will not fear me, but you will desire to please me.

Mistakes to Avoid

One mistake some people make during imprint training is not positioning themselves, the mare, and the foal safely. All parties involved in the imprinting process should be able to see one another. The foal handler should be on the ground behind the foal. The mare handler should control the mare with a halter and lead, allowing her to see and sniff her foal if she wants. But the handler should keep the mare a safe distance from both the foal and the foal handler.

The mare handler shouldn't get careless. That person's responsibility is to stand up (some people kneel down) and control the mare so that no possible injury can occur to either the foal or the foal handler. It makes no difference if the mare is gentle. I know of one case where a gentle mare pawed while the foal was being imprinted and accidentally split the foal handler's scalp.

The foal handler should fold the foal's legs and not allow them to extend where the mare can accidentally step on them, crushing a bone and crippling the foal for life.

The other mistake occurs when the foal is turned over to desensitize the other side. The danger is not to the foal, but to the handler. A newborn's legs are long and they're surprisingly strong. They have to be. In a day or so that foal will be able to keep up with his dam when fleeing from danger. Soon after birth, a foal can kick hard enough to rupture a person's spleen, or loosen a lot of teeth.

Turning the foal over is best done by two people. One handles the foal's forequarters, keeping his or her face away from the foal's front feet. The handler should firmly grasp the front legs by the cannon bones. The other person handles the hindquarters, keeping the foal's legs extended and staying far enough away from them so that a sudden kick won't do any damage. A loop of soft rope can be used to do this to make sure that nobody gets hurt. The two people should turn the foal over in unison if possible. Again, make sure the mare can see her baby in the new position.

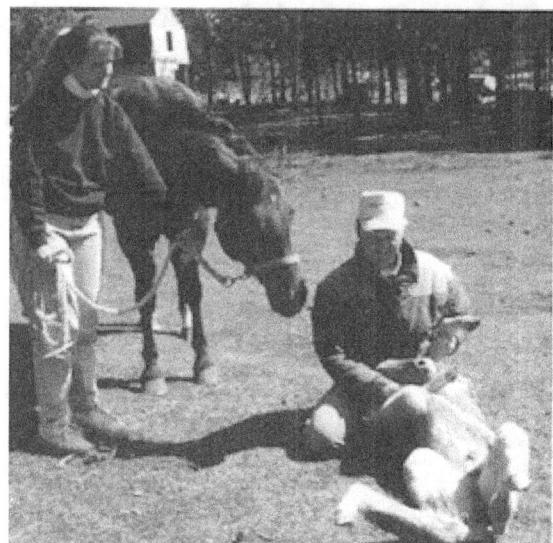

If there is no other person handy to help turn the foal, the foal handler will have to do it by himself. He should position the foal between his thighs, grasp the front legs by the can non bones, and twist the foal over. For safety, the foal handler should not have his back to the mare. She could strike or bite.

Photo by Kathy Kadash

The hind legs should follow; but if they don't, the handler will have to hold the front legs in one hand and use his free hand to push the hind legs over.

Photo by Kathy Kadash

Although it is easier if two people turn over the foal, it isn't always practical to have three people present for imprinting. Indeed, sometimes you're lucky to have two people on hand. If there are only two people, the foal handler will have to turn the foal by himself. This is best accomplished by the handler positioning the foal's forequarters between his thighs as he kneels on the ground. The handler should then grasp the front legs by the cannon bones and twist the foal's body to the opposite side. The hind legs should follow, but the handler should watch out for them.

If possible, the handler should lean back as the legs come over to prevent getting hit by them. If the hind legs don't automatically follow, the handler will have to secure the front legs in one hand, grasp the hind legs with the free hand, and pull them over.

Conclusion

It is remarkable, considering how many things can be done incorrectly, that poor results are rarely obtained when imprinting a foal. From the many reports I receive, I am sure that 99 percent of the time, imprint training has given excellent results, even when everything hasn't been done exactly right. This is because most horses have naturally submissive personalities, and super-dominant individuals are the exception. However, as in any horse training method, it must be done correctly to obtain optimum results.

I would like to mention that since the publication of *Imprint Training of the Newborn Foal,* the technique has attracted worldwide enthusiasm and acceptance. Some people are also giving me the credit for "invent ing" imprint training. This is not true, as I have clearly stated. All I did was coin the term "imprint training," because it is done during the imprinting period, right after a foal is born.

Although I have done all I can to popularize the method, I never claimed to have discovered it because I am very aware that some horsemen, throughout history, have done some training on new born foals. In fact, I have a veterinary book published in 1904 that advises, "the best time to begin educating a colt is when he is but a few hours old."

Reprinted from *Western Horseman,* March 1996.

SINCE THE publication of my book, *Imprint Training of the Newborn Foal,* in 1991, the method has been met with worldwide acceptance. I get numerous reports from horse breeders enthusiastically describing their satisfaction.

In 1995, many imprint-trained horses won races. On Thanksgiving Day, 1994, I had the satisfaction of seeing a gentle, well mannered Thoroughbred quietly enter the starting gate at Bay Meadows racetrack near San Francisco. He was the first horse out of the gate, held the lead all the way, and finished 4 lengths ahead of the second-place horse. He came from a ranch that imprint trains all of its foals.

In the introduction to this book, I state that there is historical evidence that some other cultures trained, or shaped the behavior of, newborn foals. I personally had two clients—one a warmblood trainer from Germany, the other a Paso Fino trainer from Colombia—who told me that their families had worked with newborn foals for generations, even though doing this is not traditional in either country.

Several people have told me that their Irish grandfathers taught them to handle foals right after birth, carry them around the stall, blow in their nostrils, and turn their heads from side to side. I have also read reports about certain Native Amer-

IMPRINT TRAINING

Jan Martin, broodmare manager at Flag Is Up Farm, watches as a foal learns to lead. **Photo by Debby Miller**

Now the foal learns to follow the extended right hand.

Photo by Debby Miller

i can tribes similarly handling newborn foals, but have never heard a personal verification about this practice until recently.

Harold Wadley, a resident of St. Maries, Idaho, sent me an interesting letter. With his permission, I am presenting part of that letter here:

> Dear Dr. Miller:
> What my grandpa called "taking a horse's spirit" was the same method we used for taming coyote pups we took from the den before their eyes opened.
> We each had a mare given to us for the gestation period, and especially starting with the sixth month, we brought the mares into the corrals for handling. As we curried the mares we talked especially to the side of the mare, to the colt, with the fiat of our palm pressed firmly against her stomach in different places as we talked to the colt. We did this on both sides of the mare. When the colt turned over and dropped to the bottom of her stomach, we would do this three or four times a day with one hand under her stomach and the other hand, palm down, on her back.
> When the foal arrived, we waited until the mare talked to the foal before we would say anything to it. After that, and when it was on its feet with the umbilical cord severed, we then said the same thing to the foal we had been saying all along, and put our hands around its nose to smell us.
> We were to wait then until after it nursed and rested before starting again by rubbing it all over and saying the same things to it. The words were things like what a nice colt it was, and that we were

148

waiting for it—soothing type words and never raising the voice. We were to never raise our voice to the horse later, unless it was some kind of emergency and the horse must react quickly, like a mean cow coming and the horse didn't see her yet.

For 8 to 10 days we handled the colt in the morning and late in the day by scraping its hoofs, sides and bottoms, with a stick or tap ping with a hammer. A rope was thrown over its head gently, as well as a halter; we rubbed inside its mouth, ears, and all over it.

At about 5 months or so, when weaning, we put a blindfold on the colt and walked all around it talking to it for at least 30 minutes, and then removed the blindfold and gave the foal a handful of feed. My grandpa said this took the colt back to before birth, and that we now became its provider. Even though it was scared with the blindfold, we could remove it and the colt found us there with food. After this, we never had any trouble working with the young horse.

Dr. Miller, some of this seemed to be what you were doing in the video I bought of yours, as well as what I read about the process called imprinting. My grandfather, a full-blood Cherokee, was the medicine healer in our area. My sister now is, but she doesn't have horses anymore. If this is of any value to you, I'm happy to pass it on.

Most of our Cherokee people back home have lost this, and what a pity, I think. I have heard that the peoples of the Southwest have a simi lar custom, but I have never seen it. I taught natural resource management at Haskell Indian Junior College in Lawrence, Kan., for a while, and none of the young Southwest people in attendance knew it. Some Comanche in western Oklahoma used a similar system.

Best regards in all that you do. Hank Wadley

Apparently everybody who tries imprint training, and who does it correctly, becomes an advocate of the method. There are, however, still some skeptics who cling to the traditional and dogmatic belief that handling young foals will "spoil" them. Most of these doubters and naysayers are old-time horsemen who are reluctant to change their ways.

A few academic behavioral scientists have protested that controlled scientific experiments have not been done to verify the effectiveness of what I call imprint training. In fact, several universities with equine science programs are attempting to do just that.

The problem is that, in order to scientifically validate or invalidate the method, a large number of mares would have to be bred repeatedly to the same stallion for each pregnancy; half of the foals would then be imprint trained, and the other half handled traditionally as a control group. After a decade, the results could then be statistically analyzed.

The results would then confirm what I have observed in a lifetime of experience: Newborn foals, trained as I describe in my book, consistently turn out to be gentle, well-mannered horses, who enjoy and respect humans, and who are extremely receptive to future training.

In this regard, I received a copy of another letter written to a German horse owner who had questions about imprint training.

The letter was written by Monty Roberts, owner of Flag Is Up Farms of Solvang, California. A breeder of great racing Thoroughbreds (Alleged and An AcT), and a horse training clinician, Monty has imprint-trained every foal born on his ranch for several years.

With Monty's permission, here is his interesting letter:

Dear Mr. Jacobs:
I have now received from the Jockey Club a computer readout on the mare L'Adorable, who was the object of the study with regard to foal imprinting. Please understand that I am not suggesting that we have very nervous or destructive mares. What I am suggesting is that if foal imprinting can be proven successful with very difficult mares, then it follows that it can probably be helpful with normal mares.

The subject mare is L'Adorable, foaled April 8, 1980. Her sire is L'Natural by Raise A Native, and her dam is Dors by Corporal II. It is reported that L'Adorable was an extremely difficult yearling, and while she was in the hands of competent horse people she remained difficult through the breaking process, and ultimately injured herself while in a fit of rage in the stable, kicking the side out of a box stall, causing serious injuries to her hind legs.

After a period of healing, she was placed in training and while it was difficult, she was brought along sufficiently so that it was determined that she could go to the racetrack. During transportation she once again became fractious and very nearly destroyed the trans porter. In this act she damaged her legs to the extent that it was determined by the veterinarians that she should not be raced.

She was bred in 1982 and produced a filly foal in 1983. The foal born in 1983 died in 1984, the result of a fit of rage while having her feet trimmed.

L'Adorable produced a foal in 1984 who was put down in 1986 after being deemed an incorrigible by his handlers.

She had a colt born in 1985, who was killed in 1986 in a handling accident when he bashed into fences and walls while being handled by two men who were reported to be competent horsemen and who were not related to the death of the first two.

In 1986 L'Adorable produced a colt who was in fact broke, raced, and won, but was put down in 1989 for being a very mean and aggressive 3-year-old. Again, this colt was handled by people unrelated to the handlers of the first three offspring.

In 1987 the mare had a filly who died in 1988, the result of a very strong act of aggression toward other yearlings in the field and then toward the handlers, who said that she worked herself into such a state of aggression that she simply died in a fit of rage, apparently of a heart attack.

L'Adorable gave birth to a colt in 1988, and at that time entered into an experimentation whereby this colt, later named Verdicchio, was foal-imprinted. The imprinting was done by people who were not well-trained in this new science. The imprinting procedure, as it is reported, was probably less than half as effective as we would do it today. However, this colt did live, was trained, and did win three races in North America. He is still alive and reportedly is of normal temperament.

L'Adorable produced a filly in 1989, named Tissar's Jule, who was imprinted at birth in a bit more sophisticated manner, was trained and raced in one race. Apparently she did not have strong racing ability and was retired to the show ring, where she competes now and is reported to have a pleasant disposition.

In 1990 the mare produced a filly later named Adorably, who was broke, performed well, and is alive today. She was imprinted in the first hour and in a more professional manner than the first two imprinted foals.

In 1991 L'Adorable produced a colt later named Out the Dor, who was broke, is doing well, and is reported to have a normal disposition. This colt was imprinted in the first hour and then followed up with two more imprint sessions in the first week.

In 1992 she produced a filly now named L'Ole who is in training here in California. This filly was imprinted by what I consider to be the best and the newest ideas regarding this science, and was at 18 months of age placed in training with us at Flag Is Up Farm. I can state that she was a very good student, has taken her early training well, and was sold to Kjell Zvale for racing here in California. She was working extremely well and brought $40,000. She is training in San Francisco and, at this moment, is known to have a sweet disposition by all who handle her.

In 1993 L'Adorable had a chestnut colt named Outlaw Heart, who is currently in training with us here at Flag Is Up and appears to be quite normal in all respects.

I cannot imagine a more dramatic set of facts to prove up the value of foal imprinting. This mare, after her own problems, gave rise to five foals who met violent deaths very early in their lives because of behavioral problems. Then, when foal imprinting was introduced, she gave rise to six additional foals who have lived normal existences and whose dispositions have been dramatically different from the first five.

I do not believe a more demonstrative experiment could have been conducted than this one. I have investigated the facts for you so that I am sure that they are not circumstances made up, and the mare was not in the hands at any time of anyone trying to promote science.

Sincerely,
Monty Roberts

Conclusion

I call my method of training newborn foals imprint training because it shapes behavior quickly, and lastingly, during the imprinting period of the newborn foal's life. The imprinting period is the first of the critical learning times that occur in horses during the first few days after being foaled.

Since the publication of this book on this subject, the method has become popular all over the world. The two letters presented here were extremely interesting to me and supportive of my work. I think that many other breeders using the method will find them as informative as I did.

EPILOGUE

Pat Parelli, a horse behaviorist and teacher.

In the first chapter of this book, I predicted that more competent horse trainers than I will adopt this (imprint training) method and use it to achieve things that I am not capable of achieving.

During the year that passed between the writing of this book and its publication, that prediction was fulfilled. But first, some background information.

I toured Australia the summer of 1988 with horse trainer Pat Parelli, who now lives in Pagosa Springs, Florida. Pat prefers to be called a behaviorist and a teacher rather than a horse trainer. He is all of these, and I believe is a horseman in the finest sense of the word. We put on a series of horsemanship clinics for the Aussies, and, in so doing, were heavily exposed to each other's ideas.

Pat heard me lecture repeatedly on the training of newborn foals, watched my films and demonstrations, and observed the effects of the method.

Unbeknownst to me, he started to use imprint training on his ranch and, in April 1991, I was privileged to see the results.

They are sensational. He has expanded my method and techniques into a 7-day training program starting at birth.

It involves a lot more contact between the newborn foal and humans than I have ever spent, certainly more than I talk about in this book. It employs most of the techniques I have described, with some modifications. Parelli has, as I predicted someone would, included certain performance-related factors, such as lead changes.

After the 7-day program, which he calls "an investment in the future," his foals are turned out and they receive no handling thereafter, except for that which is minimally necessary for parasite control, vaccinations, and foot trimming. When Pat's yearlings were rounded up and corraled in April '91, they were, as has been my experience, well-mannered, responsive, and gentle.

Imprint training *works,* and it works *consistently* and very, very *effectively.*

Seeing a skilled horseman use this method, with innovations and variations of his own, so effectively is very gratifying to me. It is, I believe, only the beginning of a new era in horsemanship.

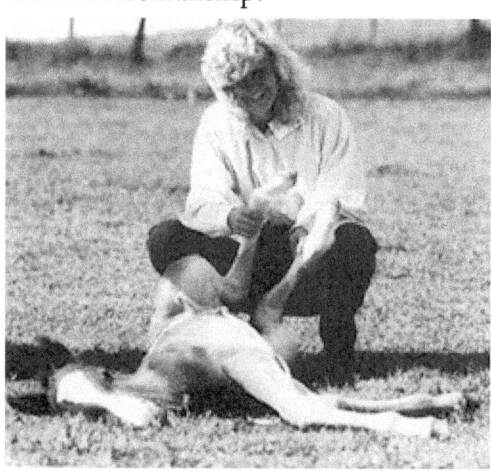

A student working with a totally relaxed, imprint trained foal at Pat's ranch.

SUMMARY

An Ideal Training Schedule for the Newborn Foal:

1. First Session—At the time of birth.
This is the imprinting period. Do the desensitization procedures described in Chapter 7. It should take about an hour.

2. Second Session—When the foal is standing. This should be within a few hours of birth. Do the desensitization procedures described in Chapter 8 and the sensitization procedures described in Chapter 9.

3. Third Session—When the foal is moving about in a coordinated manner. Usually from 12 to about 24 hours of age, more or less, depending upon the strength and coordination of the foal. You don't want to do this while the foal is weak and wobbly, but the earlier it's done, the easier it will be. Teach the foal to tie as described in Chapter 10.

4. Fourth Session—The following day.
Reinforce all procedures, especially leading and tying. Load mare and foal into a trailer.

5. Fifth Session—The following day.
Reinforce all procedures. The foal should be leading well.

6. One Week Old—Lead from the mare as described in Chapter 11. Use a loose butt rope.

7. Eight Days Old—Reinforce all procedures. Lead the mare again. Today the foal should lead at a trot. When he does, try the fixed butt rope described in Chapter 11.

8. Nine Days Old—Reinforce all procedures. Lead in hand at a walk. Teach the halt and the back in hand.

9. Ten Days Old—Reinforce again.
Lead some more. Teach the foal to stand quietly at attention as described in Chapter 12.

10. Two Weeks Old—Reinforce.
Lead in hand. Then lead from the mare and try some of performance procedures described in Chapter 13.

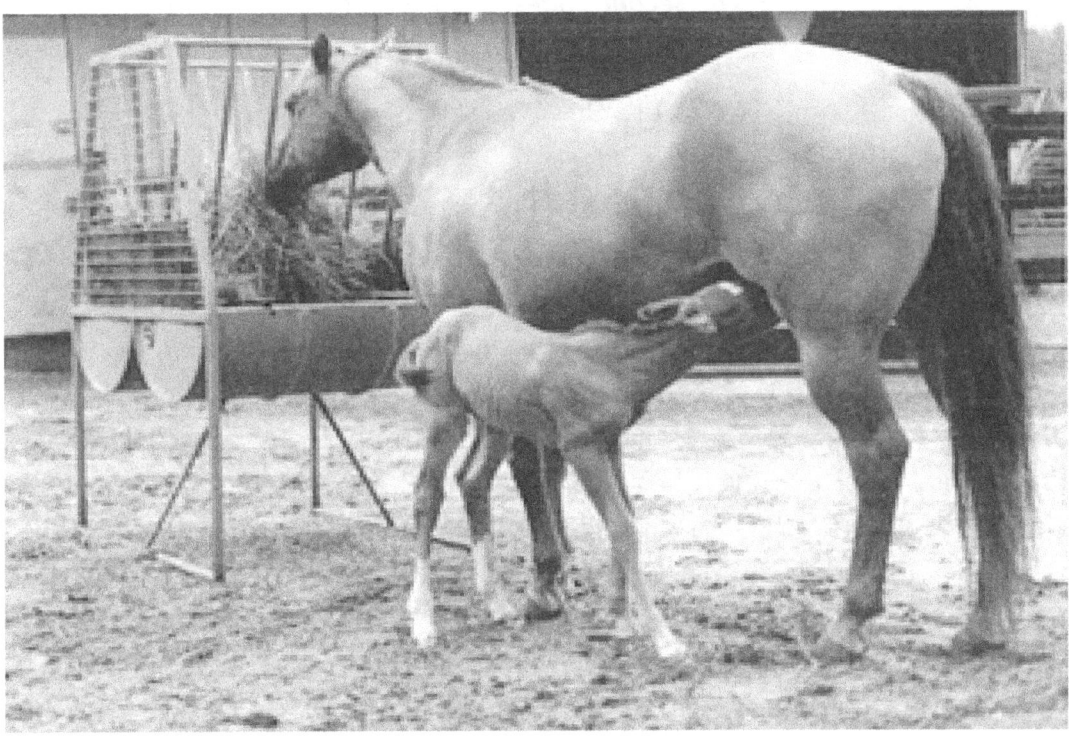

Other Works by Dr. Robert M. Miller

Books
Natural Horsemanship Explained – From Heart to Hands
The Revolution in Horsemanship (Co-authored with Rick Lamb)
Understanding the Ancient Secrets of the Horse's Mind
The Passion For Horses & Artistic Talent – An Unrecognized Connection
Mind Over Miller
Yes, We Treat Aardvarks

Equine Videos
Understanding Horses
Safer Horsemanship
Early Learning
Control of the Horse
Influencing the Horse's Mind
The Causes of Lameness

Cartoon Books
Am I Getting To Old For This?
The Second Oldest Profession
Ranchin' Ropin' an' Doctorin'
A Midstream Collection

Websites
visit www.robertmmiller.com
www.rmmcartoons.room
www.thepassionforhorses.com